Richard Sisley

Epidemic Influenza

Richard Sisley

Epidemic Influenza

ISBN/EAN: 9783743373136

Manufactured in Europe, USA, Canada, Australia, Japa

Cover: Foto ©berggeist007 / pixelio.de

Manufactured and distributed by brebook publishing software (www.brebook.com)

Richard Sisley

Epidemic Influenza

EPIDEMIC INFLUENZA:

NOTES

ON ITS

ORIGIN AND METHOD OF SPREAD.

BY

RICHARD SISLEY, M.D.,
MEMBER OF THE ROYAL COLLEGE OF PHYSICIANS OF LONDON.

LONDON
LONGMANS, GREEN, AND CO.
AND NEW YORK: 15 EAST 16th STREET
1891

All rights reserved

PRINTED BY EYRE AND SPOTTISWOODE,
Her Majesty's Printers.

CONTENTS.

CHAPTER I.
THE NOMENCLATURE OF INFLUENZA.

The Nomenclature of Disease. Clinical and Pathological Classifications. Names of influenza founded on symptoms or origin. Fancy names. Influenza, the meaning of the word. Reasons for adopting this name - - 1–5

CHAPTER II.
DEFINITION.

Definitions of Parkes, of Bristowe, and of a writer to "The Times"; the latter accepted with some necessary additions - - - - - - - 6–8

CHAPTER III.
THE ORIGIN OF INFLUENZA.

Theological views. Comets. Volcanic eruptions. Electrical and other states of the air. Flies and locusts. Influenza is prevalent in Central Asia, and endemic in China. The ultimate cause of the disease is probably a microscopic organism - - - - - 9–17

CHAPTER IV.
THE METHOD OF SPREAD OF INFLUENZA.

Views of Hirsch, Renvers, Fürbinger, Bäer, Becher, Schnirer, Colin, Bouchard. Report of Local Government Board for Ireland. The teaching of Sir Thomas Watson and of Dr. William Squire. Case of the "Stag." Watson's cases of sudden outbreaks in Portsmouth and London (1833). Reported sudden outbreaks on ships (1782). Gray's remarks on these cases - - - 18–32

CHAPTER V.
ON THE SPREAD OF INFLUENZA BY CONTAGION.

Influenza is contagious, and is chiefly, if not entirely, spread by contagion. Nature of the evidence offered in support of these views - - - - - 33–36

CHAPTER VI.

THE SPREAD OF INFLUENZA BY CONTAGION IN 1782 AND 1803.

Page

Gray's account of the epidemic of 1792. Cases of contagion mentioned by Dr. Clark, Dr. Samuel Foart Simmons, Dr. Hamilton, and Dr. Haygarth. Cases of contagion in 1803. The gradual rise and fall in the death rate in London during the epidemic of 1782 - 37–46

CHAPTER VII.

THE EPIDEMIC OF 1889–1890. BOKHARA. ST. PETERSBURGH. BERLIN.

Influenza in Bokhara in May 1889, height of epidemic in July. St. Petersburgh affected in October, height of epidemic in November. Berlin affected in November, height of epidemic in December. Professor Bäumler on contagion - - - - - - - 47–53

CHAPTER VIII.

THE EPIDEMIC OF 1889-1890. FRANCE.

Influenza recognised in Paris in November, height of epidemic in December. Influenza imported to Montpellier from Paris, and isolated cases preceded the epidemic. Imported into Frontignan from Paris; to Montbéliard from Neufchâtel and Solenre; to Vergèze from Lunel. Dr. Antony and M. Barth on contagion in hospitals. Spread by contagion on the "S. Germain." "Bretagne" infected by contagion, ships moored near escaped - - - - - - - 54–61

CHAPTER IX.

THE SPREAD OF INFLUENZA IN ENGLAND IN 1889, 1890, 1891.

Dr. Buchanan reported cases in October; Dr. Shelly in November. Epidemic in Westbourne Grove and Hammersmith in December. Outbreak at the General Post Office and at Dr. Barnado's Homes in January. Casualty patients at St. George's Hospital. Epidemic at Broadwood's Factory. Colchester, Canterbury, Chelmsford, Cardiff, and Oxford affected in January. Isolated cases existed before there was an epidemic.

Outbreak at University Press, Oxford. Isolated cases at Bristol, Birmingham, and Sheffield in January; epidemics followed. Cases imported into Liverpool from London. Nottingham affected in January, height of epidemic in February. Manchester affected in November 1889, height of epidemic in February 1890. Inverness, Stourbridge, Trowbridge, Wimborne, St. Ives, and in the rural parts of Derbyshire affected in February; Wales in March. Imported into Churchingford from Paris in December 1889, and in February from surrounding towns. Scattered cases during summer and autumn 1890, and winter of 1890-1891. Imported into Hull from New York, March 1891, spread to Sheffield and London. Cases of spread by contagion. Outbreaks at Haileybury School - - - - - 62-89

CHAPTER X.

PRISONERS AND THE INFLUENZA EPIDEMIC.

In 1803 the inmates of Ryegate Workhouse, Hereford Asylum, and Worcester Jail escaped the epidemic, and in 1889 the nuns of Charlottenbourg escaped the epidemic. At Morningside Asylum isolated cases preceded the epidemic. The prisoners escaped influenza at Bodmin, Ipswich, Kendal, Knutsford, Lewes, Norwich, and Portsmouth, although it was raging in these towns. The records of the Scottish prisons show that in none of them was there a "sudden visitation." - - - - 90-102

CHAPTER XI.

THE SPREAD OF INFLUENZA BY PARCELS.

Dr. Mease's case (1782). Cases of Dr. Dauguy des Deserts, Mr. Bernard, Dr. B. Thorne, Drs. Guiteras and White, Professor Schauta's case. The spread of influenza by parcels is not proved by any of these cases - - 103-109

CHAPTER XII.

THE INCUBATION PERIOD OF INFLUENZA.

In M. Antony's cases the period varied from one to four days; in Dr. Bordone's cases from two to five days; in Dr. Tueffart's it was four days in two cases and five days in one case. In two cases reported by Dr. Védel it was only a few hours, and in Dr. Delépine's less than twenty-four hours. Cases of a sailor on the "Correo" and of the mate on the Outer Dowsing Light-ship - 110–116

CHAPTER XIII.

INFLUENZA IN ANIMALS.

Influenza in horses. "London fever." Professor Axe. Professor Fleming on influenza in horses, dogs, and cats. Mr. Caird believes that influenza may be communicated from horses to cats and to man, M. Auguste Ollivier, that it can be communicated from man to cats, and from one cat to another, MM. Megnin and Veillon that dogs suffer from a disease analogous to, if not identical with, human influenza. The animals in the gardens of the Royal Zoological Society -   -   -   -   - 117–130

CHAPTER XIV.

The notification of influenza should be compulsory - - 131, 132

APPENDIX.

Infectious Disease (Notification) Act, 1889. [52 & 53 Vict. Ch. 72] -   -   -   -   -   - 133–141

TABLE OF ILLUSTRATIONS.

	Page
1. Chart showing the death rate in London from all causes and from "Fever" during the epidemic of influenza in 1782	45
2. Chart showing the death rate in St. Petersburgh during the influenza epidemic of 1889-1890	51
3. Chart showing the death rate in Vienna during the influenza epidemic of 1889-1890	51
4. Chart showing the death rate in Berlin during the influenza epidemic of 1889-1890	52
5. Chart showing the death rate in Paris during the influenza epidemic of 1889-1890	52
6. Chart showing the number of male patients treated for influenza in the casualty out-patient department at St. George's Hospital	68
7. Chart showing the number of female patients treated for influenza in the casualty out-patient department at St. George's Hospital	68
8. Chart showing the death rate in London during the influenza epidemic of 1889-1890	69
9. Chart showing the death rate in New York during the influenza epidemic of 1889-1890	69
10. Chart showing the number of persons absent daily from Messrs. Broadwood's Factory during the influenza epidemic of 1889-1890	70
11. Chart showing the number of persons absent daily from the University Press, Oxford, during the influenza epidemic of 1889-1890	73
12. Chart showing the number of new cases of influenza which occurred amongst 8,374 persons in Nottingham during the epidemic of 1890	78
13. Chart showing the number of cases of influenza notified to the Medical Officer of Health for Manchester during the epidemic of 1889-1890	80
14. Chart showing the number of deaths which were certified as being caused directly by influenza in Manchester during the epidemic of 1889-1890	80

	Page
15. Chart showing the number of cases of influenza which occurred at Haileybury College during the epidemic of 1890	86
16. Chart showing the number of cases of influenza which occurred at Haileybury College during the epidemic of 1891	86
17. Plan of the Norwegian barqueutine "Correo" showing the arrangement of berths and the order in which the crew were affected with influenza	114

PREFACE.

"THE contagious nature of the Influenza had,
"I thought, been sufficiently proved by many
"physicians . . . . But a contrary, and, as I
"think, a very pernicious, opinion, has lately been
"supported by physicians of great respectability;
"and authors of the highest reputation, not,
"indeed, in this, but in other enlightened nations,
"have ascribed not only this, but many other
"epidemics, even the plague itself, to a morbid
"constitution of the atmosphere, independent of
"contagion. To determine whether this doctrine
"be true or false, is of the highest importance to
"mankind. Knowledge in this instance is power.
"So far as it can be proved that a disease is
"produced by contagion, human forethought can
"prevent the mischief."

Dr. HAYGARTH. [Died 1827.]

The proverb "qui s'excuse s'accuse" applies particularly to the writing of a preface; for a book should explain itself in order to justify its existence. These notes were originally intended to form part of a treatise on influenza; they are published in their present form without delay, because I believe that the pernicious views held by "physicians of great respectability," not only in this, but in other enlightened nations, have caused and are causing a neglect of precautions against the spread of the disease. This neglect has led to much suffering, and to the loss of many lives.

I hope that these notes, however incomplete, may be sufficient to prove the contagious nature of influenza, and may lead to its being included amongst the diseases of which notification is compulsory. A private individual like myself cannot hope to obtain such an amount of information on a national epidemic as falls to the lot of a public body like the Local Government Board, whose Report will, assuredly, be of the greatest value. This Report, however, is not yet published. The mass of material collected has probably been very great, and much time is, doubtless, necessary for a full consideration of the evidence. Only those who, like myself, have had the privilege or the misfortune to classify and make a digest of many thousands of such notes can fully appreciate the labour involved in the work.

I am glad to have this opportunity of thanking Dr. Wilks, F.R.S., Professor Klein, F.R.S., Sir Joseph Fayrer, K.C.S.I., F.R.S., M. le Professeur Proust, of Paris, Signor Bacelli, of Rome, and Professor Axe, of the Royal Veterinary College, for the interesting contributions they made to a paper which I wrote for "The Universal Review" in January 1890. Without such assistance I could not have ventured to write on a subject which was then new to me, nor would my interest in the various manifestations and peculiarities of influenza have been so thoroughly aroused.

I am indebted to Professor Fleming for a most excellent article on the disease as it occurs in animals.

I have also to thank Dr. Ewart, of St. George's Hospital, and Mr. Watson, of Westbourne Park, for permission to publish particulars of cases which were under their care. I have received valuable information through the kindness of Professor Wynter Blyth, Medical Officer of Health for St. Marylebone, Dr. Collingridge, of the Port of London, Dr. Delépine, Dr. Tatham, of Manchester, Dr. Blomfield, of Exeter, Dr. Blake, of Yarmouth, Dr. Daly, of Hull, Dr. Shelly, of Hertford, Mr. E. A. Hunt, of Colchester, and Mr. W. J. Townsend Barker, of Churchingford, to whom I here gratefully acknowledge my obligation.

I have also to thank my friends, Mr. John Richmond and Mr. S. Squire Sprigge, for much help in preparing the book for the press.

11, York Street, Portman Square,
      June 29, 1891.

CHAPTER I.

THE NOMENCLATURE OF INFLUENZA.

The Nomenclature of Disease. Clinical and Pathological Classifications. Names of influenza founded on symptoms or origin. Fancy names. Influenza, the meaning of the word. Reasons for adopting this name.

There is nothing on which more difference of opinion exists amongst physicians than on the nomenclature of disease. According to some authorities the clinical aspects of a disease should determine its name. According to others the classification should be founded on pathological observation. A classification founded on either method cannot be entirely satisfactory. If a clinical division be taken, it soon becomes evident that one physician attaches importance to one set of symptoms; these, a second physician ignores, or makes light of, and insists on the importance of others which have, perhaps, been passed over by the first observer. Nor is this the only difficulty which presents itself. It is found, especially in the case of contagious fevers, that in one epidemic one symptom or set of symptoms is prominent, whilst

<small>On the nomenclature of disease.</small>

<small>Clinical classification.</small>

in a second epidemic a totally different symptom or set of symptoms may be the most obtrusive.

It follows from these considerations that the experience of one physician with regard to one epidemic of a disease cannot be sufficient to enable him to judge, from his own observations, of all the possibilities of the disease he is observing, nor to justify him in designating it after one symptom or set of symptoms.

Pathological classification. In the case of some diseases a pathological classification is impossible. Of these, influenza is one.

Nomenclature of influenza. The nomenclature of influenza is a subject on which a large book might be written, but I do not propose to do more than refer to some of the names the disease has received, and to give my reasons for adopting the one I have chosen.

Some of the many names which have been given have had reference to a symptom or to a peculiarity of the malady; some have indicated its origin; some have been playful nicknames.

(i.) *Names founded on symptoms and peculiarities.*

Amongst the symptoms and characteristics of influenza, fever, catarrh, cough, headache, and its contagious nature, have attracted the greatest amount of the attention of nosologists. Hence we find the following Latin names: catarrhus febrilis, febris catarrhosa, catarrhus epidemicus, febris catarrhalis epidemica, catarrhus a contagio, morbus catarrhalis, fluxio catarrhalis, synochus catarrhalis, febris catarrhalis, defluxio catarrhalis epidemicus, rheuma epidemicum, rheuma epidemicus, (*sic*) tussis epidemica, febris suffocativa, cephalagia contagiosa.

The French have added to the list in their language: fièvre catarrhale épidémique, catarrhe épidémique, bronchite épidémique; and the Germans have contributed the names, Blitz-Katarrh (lightning-catarrh, referring to the sudden onset), Schaffhusten, and Schaffkrankheit.

To the Erse language we owe Fuacht, Sloadan, Creatan (chest disease).

The Saxon language has supplied Pose (from gepose, heaviness), which refers to the sleepy state induced.

(ii.) *Names founded on origin of the disease.*

Influenza, as observed in Europe, has passed from one nation to another, and it has been the tendency of each nation to give it a name referring to the country from which it was imported. Some of the Jewish writers called it Kurdaikis, from its supposed origin from the Kurds. To the Russians it has been Chinese catarrh, to the Germans and Italians, Russian fever. At different times in France it has been Italian fever, Spanish catarrh, and Russian influenza. During the late epidemic the latter term was applied to the disease by Continental authorities of all nationalities, and by some English writers. I see no more reason to perpetuate a national discourtesy in this case than in that of any other contagious malady.

(iii.) *Names founded on fancy.*

The imagination of the French has supplied a number of popular names, and amongst them coqueluche, horion, tac, baraquette, ladendo, grippette, petite poste, petit courrier, follette (the frolicsome), coquette, grenade, générale, ailure, cocote.

The Germans have added to the list Mödefieber, Ziep, Hühner-weh.

The name influenza, is of Italian origin, and according to Dr. Frank Clemow,* was used as early as 1580 by Pietro Buoninsegni to describe an epidemic which occurred in Florence in 1386. Influenza means an influence, and some Italian physicians ascribed that influence to the stars. The name came to England at least as early as 1743,† but did not at once get into general use, for the word influenza does not occur in the folio edition of Johnson's Dictionary (1765).

I have by no means exhausted the synonyms of influenza, but it might be supposed that there is sufficient choice even amongst those given. It appears, however, that this is not the case. At the present time (June 1891) an attempt is being made to add to the list. I can see no good object

---

* Public Health, Vol. II., p. 361. Epidemic Influenza, a paper read before the Medical Officers of Health, February 14, 1890, by Frank G. Clemow, M.B., C.M., Physician to the English Hospital at Cronstadt.

† *Gentleman's Magazine.*—It is stated in Ziemssen's Cyclopædia of Practical Medicine, that according to Biermer "Pringle first designated the disease by this name influenza" in his book "Observations of the Army," 1752. Biermer's statement is not correct, but is still current amongst medical writers.

in this. When a new name is given to an old disease, as Dr. Wilks points out, "the only advantage is to the man who names it."*

In the present book I adopt the name influenza, because—

<small>Reasons for the adoption of the name influenza.</small>

(1.) The name is well known.
(2.) It is distinctive; no other disease is called by that name.
(3.) It implies no theory, either as to origin, causation, or characteristics.

In short, to quote again from Dr. Wilks,* " Influenza covers a variety of cases which I see, and no name I could invent would similarly include them all."

---

* Private letter to the Author.

CHAPTER II.

DEFINITION.

A perfect definition of influenza impossible. Some of the best writers on the subject have not attempted to make one. Definitions by Parkes, by Bristowe, and by a writer to " The Times." That of the latter accepted with some necessary additions.

<small>A perfect definition of influenza impossible.

Some of the best writers on the subject have not attempted to make one.</small>

The greatest of our race has suggested that to define true madness is little less than mad. To sum up accurately all the symptoms of influenza in a single sentence is impossible. Some of the best writers have not attempted to do so. Of these the late Dr. Theophilus Thompson[*] is one. Dr. Thompson's classical compilation is one from which succeeding writers on the subject have freely drawn, and often without any acknowledgment of the source of their information or of their authority. No physician was better qualified to define the disease, yet Dr. Thompson wisely refrained, and began his book by speaking of "the malady which forms the subject of this volume."

---

[*] Annals of Influenza or Epidemic Catarrhal Fever in Great Britain from 1510 to 1837. Prepared and edited by Theophilus Thompson, M.D., F.R.S. London, 1852.

Many of the older writers were equally wise, or equally cautious. The works of Short, of Pearson, and of Huxham, will be consulted in vain for anything like a definition.

One of the best accounts of influenza is that given by the late Dr. Edward A. Parkes* in Reynolds's System of Medicine. Dr. Parkes's definition is as follows:— *Definition of—*
1. Parkes.

"An epidemic specific fever, with special and early implication of the naso-laryngo-bronchial mucous membrane; duration definite of fever, four to eight days; one attack not preservative in future epidemics."

This definition is not satisfactory. In the first place, "special and early implication of the naso-laryngo-bronchial mucous membrane" is by no means constant. In the second place, the duration of the fever is variable.

Dr. Bristowe's† definition is as follows:—

"A contagious catarrhal affection of the respiratory tract, of short duration, but attended with much prostration, and occurring for the most part in widespread epidemics." *2. Bristowe.*

This description is open to much less criticism. The only great objection to it is the inclusion of catarrh of the respiratory tract as a constant symptom. The contagious nature of the disorder and the prostration it causes are both mentioned.

---

* A System of Medicine, edited by J. Russell Reynolds, M.D., F.R.C.P. London, 1866, Vol. I., p. 27–50.

† A Treatise on the Theory and Practice of Medicine, by John Syer Bristowe, 2nd edition. London, 1878, p. 140.

3. An anonymous writer to "The Times."

The following definition was given by an anonymous writer in "The Times" of April 12th, 1890:—

"Influenza is a specific fever, epidemic and often pandemic, of sudden onset and short duration, attended with loss of appetite and very great prostration, associated often with more or less severe catarrh, neuralgic pains, or gastro-intestinal disturbance, and especially liable to be complicated by severe respiratory affections to which the mortality of the disease is chiefly due."

There is one important omission in this definition. The fact that influenza is contagious is not stated. Exception must also be taken to the phrase "neuralgic pain." Some of the pains felt in influenza cannot be justly described as "neuralgic," they are muscular. Otherwise the description is good, as it serves to mark off the disease from all others. I am content to adopt it with additions which I have named. These additions are essential.

CHAPTER III.

THE ORIGIN OF INFLUENZA.

Theological views. Comets. Volcanic eruptions. Electrical and other states of the air. Flies and locusts. Influenza is prevalent in Central Asia, and endemic in China. The ultimate cause of the disease is probably a microscopic organism.

The views held by men about the origin and spread of disease have always been modified by the spirit of the age in which they have lived. It is not, therefore, remarkable that influenza has excited many strange and contradictory ideas as to its cause.

Formerly, when theological views largely predominated, it was supposed that the Deity, or a false god, emitted or caused some occult power which produced the pest. A somewhat similar idea has even lately been expressed, namely, that the air was poisoned by Satan. I do not think that in the present day it is necessary to combat such views, for they are not now generally held by theologians; and I am pleased to learn that the germ theory of disease is found to be compatible with the teaching of the Jewish Prophets.*

Some writers have connected the origin of the disease with the influence of comets, which have on

---

* The Influenza Epidemic; a Visitation from God, by the Rev. Daniel B. Hankin. Churchman's Magazine, June, 1891.

more than one occasion appeared about the time at which an epidemic was raging, but no causal relation has been made out between these heavenly and earthly visitations. The emanations from volcanic eruptions have also been suggested as causes, and one observer* has gone so far as to indicate the exact chemical compound which he thought caused the malady.

Comets.
Volcanic eruptions.

Electricity. Electrical conditions of the air and of the earth have also been credited with similar power, but there is no proven basis of fact to these conjectures.

States of the air. States of the air have been largely discussed as to their causal relation with influenza, but epidemics have occurred in hot and cold countries, in damp and dry climates, and under most various barometric conditions. The presence of an excess of ozone in the atmosphere has been discussed, and, even as late as 1889, it was suggested that allotropic oxygen formed in Paris at the exhibition by the electrical machines, had been the cause of the epidemic in that city. With as little reason flies, caterpillars, and flights of locusts have been credited with an identical power for evil.

Flies and locusts.

Such views are rather of historic than practical interest, and for information on these questions the reader is referred to the work of Fleming,† and of the late Dr. Theophilus Thompson.‡

---

* Dr. Prout suggested in his Bridgewater Treatise that Influenza might be caused by seleniureted hydrogen.

† Animal Plagues, their History, Nature, and Prevention, by George Fleming, F.R.G.S., &c. London, 1871. Hall and Chapman. Animal Plagues, their History, Nature, and Prevention, Vol. II., by George Fleming, F.R.C.V.S., F.R.G.S. London, 1882. Ballière, Tindall, and Cox.

‡ Op. cit.

It may be that M. Teissier, the eminent French physician, whose views are set forth below, was puzzled by these conflicting sources of origin, and thus driven into complete scepticism, for he implies that the origin of influenza is not known. He says this, not indeed in so few words, but still with emphasis, and expresses himself thus :—

"Influenza, without cradle of origin, born no one knows where, passing like a cloud which obeys the caprice of the night, traverses at the same time or in the course of a few days the distance between towns situated at the four corners of the earth."* *Origin unknown.*

Now the statement about the rapid spread of influenza is, as we shall see, certainly not true with regard to the epidemic of 1889–1890; nor is the whole description founded on fact.

I have quoted M. Teissier's views, however, to show that the question of the origin of influenza is one that presented a good deal of difficulty to a skilled and eminent physician. The "cradle of origin" of the disease is in the east, and its birth, though mysterious, is not miraculous.

Mr. Naphtali Herz Imber† states that epidemics of influenza are mentioned by early Hebrew writers, and that some of the Rabbis call it Kurdaikis, and he adds that two varieties of the disease are spoken of, the mild and the strong. Without attaching too much importance to Rabbinical teaching, it is interesting to note that early Jewish writers traced the disorder to Central Asia. On this matter *Traced to the Kurds.*

---

* Bulletin Médical, 1890, p. 45.
† Jewish Standard, January 10th, 1890.

it is likely that Chinese records will throw some light, and there is evidence that influenza is endemic in parts of the Celestial Empire.

Mr. Gilmour,* a missionary who travelled in China, relates facts which bear on this point. His book does not profess to be a scientific one, but the information given is worth quoting. Mr. Gilmour says—

<small>Prevalent in Mongolia.</small>
"In one district (between Kiakhta and Urga) we had to ride a long stretch of many miles without entering a tent. As often as we drew up at a tent a woman or a man would come out and say, 'Dismount at my tent another time; we have the cough.' This cough seemed to be a kind of influenza much dreaded by the Mongols. As far as I can learn it seldom proves fatal, but travellers are careful to avoid it, and no one would think of using the pot and ladle of a family suffering from this sickness."

<small>Endemic in China.</small>
The only thoroughly reliable scientific evidence I have hitherto found as to the origin of influenza in China is derived from the Medical Reports of the Chinese Customs service.† An account is there given of an outbreak of influenza which occurred at Swatow, where it is stated sporadic cases of the disease occur. The report is as follows:—

"At Swatow, during the summer of 1879, a somewhat peculiar epidemic of influenza appeared. It attacked the children living on one side of the

---

* Among the Mongols, p. 65. Religious Tract Society.

† An Epitome of the Report of the Medical Officers to the Chinese Imperial Maritime Customs Service from 1871 to 1882. Compiled and arranged by Surgeon-General C. A. Gordon, M.D., C.B. London, 1884.

river, those on the opposite side, and on Double Island, being unaffected by it. It first occurred at a particular house, in a child 20 months old; its symptoms, 'running at the eyes and nose, feverish-'ness, general *malaise* and loss of appetite; at the 'end of five days a sharp bronchitic attack, then 'lessened fever, gradual resolution, and recovery in 'about 10 days.' The second attack occurred in a boy five years of age; it began in the same way, ending in a sharp attack of laryngismus stridulus and bronchitis, recovery taking place after 10 days. The disease attacked all the children living on the south side of the river, and affected all in the same way as here described. It only attacked children; and the reporter states that he had previously seen nothing similar to it, although sporadic cases of influenza are not uncommon."

This information is, I think, of the greatest importance to those who seriously wish to investigate the origin of the disease. The fact here recorded that sporadic cases of influenza arise at Swatow deserves more attention than it has yet received, and I should like to suggest to bacteriologists that a careful study of the life history of the micro-organisms indigenous at Swatow would not only open to them a new and interesting field of research, but might also throw light on an obscure subject.

At present it cannot be said that the actual cause of influenza has been found. Strong arguments are in favour of the theory that the *causa causans* is a microscopic organism. But the organism has yet to be found. It would be both useless and tedious

<small>Probably caused by a microscopic organism.</small>

Probably caused by a microscopic organism.

to record here the observations which have been made by bacteriologists, for their search has not been successful, and a record of their failures is unnecessary. Yet they have not been idle. No one could open a daily paper for some time without seeing the names of certain Austrian observers, a record of whose work was telegraphed all over the world. Scientific experiments thus advertised before they are confirmed only bring science into ridicule and contempt, a fact which has lately been abundantly proved in the case of a distinguished German savant whose previous work had commanded universal respect.

The arguments in favour of the view that the ultimate cause of influenza is a microscopic organism, were clearly stated by Professor Klein* in a letter which he kindly wrote to me in January 1890, and from which I quote the following passage :—

" That a microbe must be the primary cause of the disease is suggested by the epidemic character and by the infectiousness of the disorder. No other theory, as, for instance, that of peculiar atmospheric conditions, can for a moment be considered as compatible with the fundamental and well ascertained fact that the disease is in a high degree an infectious disease. The course of the epidemic in the various Continental cities and in London leaves no doubt about this point. In all epidemic diseases the spread from person to person cannot be explained by any but a living and self-multiplying

---

* Universal Review, January 1890.

essence. The particular specific microbes find entrance into the system of one or more susceptible bodies; herein they multiply and set up the particular disease. The infected body becomes, as it were, the soil on which the new crops of the microbe are raised, and thereby assumes the character and power of foci or centres, from which infective matter, *i.e.*, the new crop of microbes, becomes disseminated, and ready and capable of further infection or invasion of new bodies. A non-living material cannot fulfil this condition, so characteristic of infectious diseases, viz., to spread the disease from individual to individual, although a particular state of the atmosphere *might*, considered from a purely theoretical point of view, produce simultaneously in a number of persons a diseased condition; but these persons themselves could never become foci of infection. Now, this is precisely the fundamental fact which obtains in this as in other epidemic diseases. It has been observed in all Continental cities—I have had myself excellent opportunities of observing it recently in London —that in a particular establishment, a household or a school, some individual becomes smitten with the disease, then day by day numbers of new victims are gradually added to the sick list. What makes the present epidemic remarkable is its extreme infectiveness, the rapidity with which it spreads, and the susceptibility of a vast number of people towards it. These facts suggest that the microbe is one which multiplies very rapidly, that it is conveyed, and that it enters by the breath (*i.e.*, that it

Probably due to a microscopic organism.

spreads by the air), and lastly, that it finds in most persons a suitable nidus for living and thriving."

It will be noticed that the arguments used are derived from analogy. Influenza is infectious: infectious diseases are due to living organisms: therefore influenza is due to a living organism. Flint* uses the same argument, and says:—

"Being an infectious disease, the parasitic doctrine not only offers a rational view of its etiology, but reasoning from analogy is a logical inference."

Men of science are not dogmatic on any belief which is arrived at by analogy, however strong the analogy may be. On the other hand, in the absence of definite and absolute proof, it is not right to ignore any facts which may help us to see what is the most likely explanation of an obscure phenomenon.

Of the cause of influenza we may say with confidence that, although it is not conclusively proved to be due to a microscopic organism, there are very strong arguments in favour of that hypothesis. How strong these arguments are can only be appreciated by those who are acquainted with the facts which are known about the diseases which have been proved to be of parasitic origin, and the facts which are known about the life history of those organisms.

Bacteriology is a comparatively new science, one which requires for its study not only time, skill, and patience, but an elaborate apparatus. For these

---

* A Treatise on the Principles and Practice of Medicine, by Austin Flint, M.D., LL.D., revised 6th edition. London, 1886.

reasons the investigation of the ultimate causes of disease cannot be carried on by those engaged in active practice. From this it unfortunately follows that those who have the best opportunities for observing disease have the least chance of studying its cause. A division is thus formed between many of those who study the practical and those who study the theoretical side of medicine. My reason for mentioning this fact is because I feel that its effects are far reaching and disastrous. Far reaching, because it must affect a large proportion of the medical men who have been in practice for any considerable length of time, and disastrous because it prevents them from following the developments of the medicine of the future, and even from being able to thoroughly understand the nature of the evidence brought forward by bacteriologists.

It is only fair to those who have given up or never taken to a study of the developments of the germ theory of disease, to confess that much of the literature of the subject is inconclusive and uninteresting; but this is far from being invariably the case. Whoever will begin by reading the theory of the subject in Sir Henry Holland's "Medical Notes and Reflections," and study the modern developments of the new science in the works of Klein and Crookshank, will have no reason to complain that his time is wasted, and will assuredly be led to acknowledge the importance of the subject, and the great future which it opens to medicine.

CHAPTER IV.

THE METHOD OF SPREAD OF INFLUENZA.

Influenza and Catarrh not identical. How is influenza spread? Views of Hirsch, Renvers, Furbinger, Baer, Becher, Schnirer, Colin, Bouchard. Report of Local Government Board for Ireland. The teaching of Sir Thomas Watson and of Dr. William Squire. Cases in favour of the theory of aerial contamination as a method of spread. Case of the "Stag." Watson's cases of sudden outbreaks in Portsmouth and London in 1833. Cases of sudden outbreaks on ships in 1782. Gray's remarks on these cases.

It is thought by a few physicians that influenza is nothing more nor less than a modified form of common catarrh. It has also been said that cases of the disease are endemic in one London suburb.

Influenza and catarrh not identical.
Dr. Gordon Hogg,* of Chiswick, in a letter to the "British Medical Journal," says :—" I believe we have here, as elsewhere, ever endemic, a mild form of the old influenza." No confirmation of this statement has reached me from other doctors, but it is a matter of common observation that catarrhal febrile attacks are specially prevalent in

---

* British Medical Journal, December 21, 1889.

low-lying and damp neighbourhoods. I am not in a position to deny that influenza is endemic at Chiswick, but I may here point out that I do not consider that catarrh and influenza are identical.* It is not, however, seriously contended by anyone that a severe epidemic of influenza has ever broken out in England without having previously affected other countries.

How is the spread of influenza to be explained? Is there an aerial contamination? If so, how is it produced? Or is the poisonous matter given off by the bodies of patients who have the disease? To put the question briefly, is influenza spread by the air, or by man? Or is it spread in both ways? Or does it arise *de novo?* {How is influenza spread?}

These questions have given rise to much discussion, and it is necessary to see how they have been answered, and what reasons have been given for the different opinions which have been formed on them.

---

* I have lately heard a very eminent medical authority express the view that influenza and catarrh are identical. The best answer I can give to this has already been given by Dr. Gray, and I am able to adopt his words, which were written about the epidemic of 1782, and apply them to the epidemic of 1889-1890.

"At the first appearance of the disorder, before its character had been attentively observed, or its progress traced, it was in some measure excusable to ascribe it to causes which, however inadequate to the effect, were the only ones that presented themselves to the imagination; but at present, when its character has been so well ascertained, and its progress from Russia to England has been so clearly traced . . . . and is so generally acknowledged, it would be superfluous to endeavour to prove that which every one admits; or should any one think arguments still wanting to show that the late influenza was not a common catarrh, produced by the changes in the weather, he will find many of that sort in the subsequent part of this account." Medical Communications, Vol. I. London, 1784.

METHOD OF SPREAD.

Views of Hirsch.

Dr. August Hirsch is justly considered to be a great authority on this subject, to which he has undoubtedly given great attention.

The bibliography of the subject alone occupies nearly 13 pages in the English translation of his handbook.* A work of such labour must be read with respect, and the conclusions arrived at by the writer cannot possibly be ignored; I must therefore state them in detail.

"Influenza," Hirsch says, " is a specific infective disease .... and has at all times and in all places borne a stamp of uniformity in its configuration and in its course such as almost no other infective disease has. Its genesis pre-supposes therefore a uniform and specific cause ..... There can be no objection to calling this specific cause by the name of 'miasma,' so long as we remember that nothing more is expressed thereby than that which the physicians of the sixteenth and seventeenth centuries called a 'fouling of the air,' and that, in setting up a name in the place of an obscure conception, we do not bring ourselves by that means a single step nearer to a knowledge of the cause of the disease. All the opinions that have been put forward as to the nature of this 'influenza miasma' are without any basis of fact, and that is true more especially of the theory, maintained as early as the eighteenth century, and lately revived, of a '*miasma vivum*' or an organic (animal or vegetable) morbid poison, upon the

---

* Handbook of Geographical and Historical Pathology, by Dr. August Hirsch, Professor of Medicine in the University of Berlin. Translated from the 2nd German edition, by Charles Creighton, M.D. London, 1883. The New Sydenham Society, Vol. CVI.

carrying of which by the air the spread of the disease was thought to depend. But, as we have already seen, there is not the slightest cogent reason for supposing that the several parts of an influenza pandemic stand in a genetic relation to one another, or that it is a question of the conveyance of a disease-producing substance from place to place. We might with just as much probability assume that the cause of the disease has sprung up *de novo* at all places where its effects have been manifested, as that it has been distributed by the movement of the air. And, indeed, the circumstances that the progress of the disease does not depend on the direction of the wind, and may sometimes even go contrary to it, speaks in favour of the former view."

It is evident from this that Hirsch absolutely disbelieves in the hypothesis that the disease is due to any low form of animal or of vegetable life, but that he is not equally opposed to the view that the disease may arrise *de novo*. On the first question it has been admitted that it is not proved to demonstration that influenza is due to a microscopic organism, for no such organism has yet been found. There are, however, strong arguments in favour of the hypothesis that the disease is due to a yet undiscovered organism. These I have already stated. With regard to the origin of the disease *de novo* the evidence brought forward is not sufficient to make it worth while to deal with the question.

On the "alleged contagiousness" of influenza Hirsch writes:—

"The question whether influenza is communicable or contagious has given occasion to a lively controversy. In more recent times the great majority of observers have answered it decidedly in the negative, not so much on the strength of the many single observations which tell against the communicability of the disease, as on the ground that the spread of influenza can be shown to have taken place quite independently of intercourse. To this argument I may add the fact that it has not spread more quickly in our own times, with their multiplied and perfected ways and means of communication, than in former decades or centuries. 'The simple fact is to be recollected,' says Jones,[*] ' that this epidemic affects a whole region in the space of a week, nay, a whole continent as large as North America, together with all the West Indies, in the course of a few weeks, while the inhabitants could not within so short a time have had any communication or intercourse whatever across such a vast extent of country. This fact alone is sufficient to put all idea of its being propagated by contagion from one individual to another out of the question.' "

I must pause here to at once protest against this assumption of Jones. *The fact, if it be a fact, that the epidemic of influenza affected a whole continent in a week proves that that continent was quickly affected by that disease. It does not prove that the disease cannot be propagated by contagion.* I must here say that striking "facts," like the one quoted by Jones, have frequently turned out on careful investigation to be due to the fancy of the savant and not the result of observation.

---

[*] Philadelphia Journal of Med. and Phys. Sc., 1826, n.s., LV. 5.

But I do not wish at this place to destroy the continuity of Hirsch's argument.

He continues, " Partisans for the spread of influenza by contagion have found support for their views in the breaking out of the disease at various places, somewhat removed from the track of commerce, after the arrival of strangers; for example, the Danish physicians in Iceland and the Faroe Islands have found evidence of that kind in the outbreaks of influenza that have followed the arrival of foreign ships. Without questioning the accuracy of the observation itself, we may hesitate to accept the conclusions drawn from it when we duly keep in mind that the suspected importers of the morbid poison remain, as we are expressly told, unaffected by it, that they continue untouched by the epidemic, and, further, that the disease has not unfrequently appeared in these and other islands at the time of the ship's arrival, although influenza had not been prevailing as an epidemic anywhere else, and most certainly not in those countries from which the ships had sailed. These considerations, taken along with peculiarities in the incidence and course of influenza epidemics—their occurrence suddenly and without prelude, and their attacking the people *en masse*, their equally sudden and complete extinction after a brief existence, generally of two to four weeks, and the frequent restriction of the disease to *one* place, while the whole country round has been completely free from it—all these points are so foreign to the mode of development and the mode of spreading proper to such maladies as originate beyond doubt through the communication of a morbid poison, that we shall find it hard to discover any reason for counting influenza among the contagious or communicable diseases."

It is worthy of notice that Professor Hirsch adopts the teaching of Jones, and argues that we shall find it hard to count influenza "among the contagious or communicable diseases," because he says that certain points in its spread are foreign to the "mode of development and the mode of spreading proper to such maladies as originate beyond doubt through the communication of a morbid poison."

In short, the teaching of Hirsch is that influenza is not contagious. I do not know whether extended experience has confirmed or altered Hirsch's views, The quotations I have made are from the translation of his book published in 1883, which is still a standard book in this country.

Views of Renvers, Fürbinger, Bäer, Leyden, Schnirer.

Other German physicians have lately expressed views similar to those of Hirsch. Thus, Renvers has taught the theory that it is due to a general poison which is born at the same time and dies at the same time at different points of the earth's surface.

Fürbinger states that influenza is not contagious, and Bäer maintains that it is absurd to talk of such a thing, and Leyden quotes with approval an expression of Becher comparing the epidemic to the box of Pandora.

Amongst the Austrians, Schnirer denies the contagious nature of influenza. I give these opinions for what they are worth, and on the authority of Professor Grasset, of Montpellier,* to whose excellent monograph I am greatly indebted.

Modern French physicians have freely expressed their views on the spread of influenza, and in many cases they have drawn on their imagination for their facts.

Thus, in France, M. Colin has delivered himself after this manner :—

"The present epidemic shows once more that it is independent of all transport by human means, for it spreads with equal quickness over seas, and through uninhabited regions, as it does through a densely populated country. Only light and electricity, the most rapid physical agents, travel as quickly."

This statement, as I shall show, is not accurate, but upon it is founded M. Colin's belief.

"I believe that it follows," he says, "from my communication, that influenza is a disease which depends on external atmospheric conditions, and that it is independent of human contamination. I do not believe that influenza is contagious."

Dr. Bouchard has spoken in similar terms :—

"Influenza attacks, it is true, a great number of individuals, but that is a proof that the disease is not contagious. A contagious malady would not attack in a single night 50,000 people, a thing which happened in 1858. Three weeks ago the

---

* Leçons sur la Grippe de l'Hiver, 1889–1890. Montpellier, 1890.

disease was recognised at St. Petersburg, and its method of spread shows that it has followed the lines of human intercourse."

It is only fair to Dr. Bouchard to add that extended experience convinced him that the complications of influenza were infectious even if the original disease was not.

The Report of the Local Government Board for Ireland. Continental authorities are not alone in believing that contagion plays no part in the spread of influenza. The last report of the Local Government Board for Ireland contains a rather brief account of the epidemic of 1889–1890. The report does not show evidence of much care in its preparation, but the following paragraph is of interest:

"On the subject of the origin and mode of extension of the epidemic in Ireland, the views expressed by medical officers differ, but there is a general consensus of opinion to the effect that the disease was of a miasmatic character, that it was air-borne, that it was preceded and accompanied by high temperature and moist atmosphere. A small number of observers considered it infectious and contagious, but the great majority were of opinion that it was not."

The teaching of Sir Thomas Watson. The teaching of Sir Thomas Watson[*] has probably affected living physicians more than has the teaching of Hirsch, for most of us have consciously or unconsciously been affected by the writings of the most graceful and convincing medical writer of the century. Watson was much in advance of his time in appreciating the importance of the germ theory of disease, a theory lucidly expounded by

---

[*] Lectures on the Principles and Practice of Physic, by Thomas Watson, M.D. London, 1843, Vol. II.

his great contemporary, Sir Henry Holland,* whose classical book "Medical Notes and Reflections" has of late fallen into undeserved neglect. Watson saw clearly enough what Hirsch has failed to grasp, that even if it could be proved that influenza was chiefly spread by an aërial contamination, this did not disprove that it could be spread also by contagion. Watson says distinctly, "I will not say that the disease may not be in some degree infectious, for there is reason to believe that other epidemic disorders, having many points of analogy with the influenza, *are* somehow imparted from one individual to another, although they are mainly produced by some influence which resides in the atmosphere. There are points in the history of influenza which furnish a strong presumption that the exciting cause of the disorder is material, not a mere quality of the atmosphere, and that it is at least *portable*. The instances are very numerous, too numerous to be attributed to mere chance, in which the complaint has first broken out in those particular houses of a town at which travellers have recently arrived from infected places."

But although the knowledge and candour of this great teacher led him to make this statement, it is quite clear that he believed that the spread of influenza was due to a poisoned condition of the atmosphere. On this point he says,—

"I have remarked that Cullen makes this species of catarrh to proceed from contagion. But the

---

* Medical Notes and Reflections, by Henry Holland, M.D., F.R.S. London, 1839.

visitation is a great deal too sudden and too widely spread to be capable of explanation in that way."

Dr. William Squire's views.

Amongst living English writers, Dr. William Squire* apparently attaches even less importance to the spread of influenza by contagion than did Watson. In an article contributed to the "Lancet" on the subject, after a review of the evidence he recorded, Dr. Squire sums up and says:—

"The balance seems to be rather against direct personal infection as a frequent or potent cause of the spread of influenza, but sufficiently possible to enforce a caution against introducing a doubtful visitor amongst the weakly or infirm."

Cases in favour of aerial contaminations as a method of spread.

Some physicians have declined to consider whether influenza is spread by contagion because instances have been recorded in which it has appeared that the infection must have been derived from the air. These instances may be conveniently arranged under two heads:

(1.) Cases in which large numbers of the population have been suddenly seized.

(2.) Cases in which influenza has broken out in ships at sea.

Sir Thomas Watson† says:—

Case of the "Stag," and Watson's cases of sudden outbreaks in London and Portsmouth in 1833.

"On the 3rd of April in that year (1833), the very day on which I saw the first two cases that I did see of the influenza, all London being smitten with it on that and the following day—on that same day the 'Stag' was coming up the Channel, and arrived at two o'clock off Berry Head, on the

---

* The Infection of Epidemic Influenza. Lancet, April 19, 1890.

† Op. cit.

Devonshire coast, all on board being at that time well. In half an hour afterwards, the breeze being easterly and blowing off the land, 40 men were down with the influenza, by six o'clock the number was increased to 60, and by two o'clock the next day to 160. On the selfsame evening a regiment on duty at Portsmouth was in a perfectly healthy state, and by the next morning so many of the soldiers of that regiment were affected by the influenza that the garrison duty could not be performed by it."

Watson's account of these outbreaks is striking and almost convincing. But with regard to the sudden outbreak in London, it must be said that Watson could not by any possibility have been able to get complete proof that there were no cases of influenza in the town before April 3rd. I do not think, therefore, that the statement should be considered as a statement of fact, but as a statement of opinion founded on insufficient data. Similar statements of sudden visitation of influenza were made by some Continental observers during the late epidemic, but careful inquiry proved that in all instances isolated cases of the disease preceded a general infection of the population.

The account of the outbreak at Portsmouth, although related to Watson " on good authority," is no more reliable than his own account of the London visitation, for there could have been no higher authority than that of Sir Thomas Watson himself.

The case of the "Stag" is frequently quoted, and it is not the only one of the kind on record.

Cases of sudden outbreaks in

Ships in 1782.

The Medical Transactions published by the College of Physicians in London* contain the following records.

"On the 2nd of May 1782 the late Admiral Kempenfelt sailed from Spithead with a squadron of ships under his command, of which the 'Goliah' was one, whose crew was attacked with the influenza on the 29th of that month; the rest were affected at different times; and so many of the men were rendered incapable of duty by this prevailing sickness, that the whole squadron was obliged to return into port about the second week in June, not having had communication with any shore, and having solely cruised between Brest and the Lizard.

"About the 6th of May, Lord Howe sailed for the Dutch coast with a large fleet under his command; all were in perfect health; towards the end of May the disorder first appeared in the 'Rippon,' and in two days after in the 'Princess Amelia.' Other ships of the same fleet were affected with it at different periods; some, indeed, not until their return to Portsmouth about the second week in June. This fleet also had no communication with the shore until their return to the Downs, on their way back to Portsmouth, towards the 3rd or 4th of June."

On page 60 of the same volume the following case is given:—" Information has been received, that this distemper broke out and became very general among the crew of the 'Altas' East Indiaman in September 1780, while that ship was

---

* Vol. III., p. 62.

sailing from Malacca to Canton. When the ship left Malacca there was no epidemic disease in the place; when it arrived at Canton, it was found that at the very time when they had the influenza on board the 'Atlas' in the China seas, it had raged at Canton with as much violence as it did in London in June 1782, and with the very same symptoms; but with the addition of bilious complaint, which also accompanied its appearance on the coast of Coromandel, and in Bengal, where it raged nearly about the same time."

These cases are so frequently quoted or referred to that they appear to possess almost the authority of an inspired medical writing, if such a thing can be imagined. Dr. Edward Gray* refers to these cases, and comments on them. Gray's article "An Account of the Epidemic Catarrh of the year 1782" was compiled at the request of a Society for Promoting Medical Knowledge, and is a monument of careful research and just reasoning. I unreservedly accept all he says on this subject, and am convinced that the question could not be considered in a clearer light. I therefore adopt his words as well as his reasoning.

Gray's remarks on the cases recorded in 1782.

"It is credibly affirmed,"† he says, "that the crews of several ships were seized with the influenza many miles distant from land, and came into various ports of England labouring under it; the same thing is said to have happened to ships in the East Indies, and other parts. A want of

---

* Medical Communications, Vol. I. London, 1784.
† Op. cit.

precision, or of authentication respecting the circumstances above alluded to, makes it improper to draw any inferences from them; but without pretending to deny the truth of them, the following anecdote will serve to show that great caution is requisite before they are admitted.

"Mr. Henry, of Manchester, informed the society, from what he thought good authority, that a ship from the West Indies to Liverpool was, by stress of weather, driven out of her proper course into a higher north latitude, where her whole crew were seized with influenza; but wishing afterwards for more accurate information on the subject, he wrote to Dr. Currie, of Liverpool, desiring him to make every necessary inquiry into the matter; that gentleman, who took great pains to investigate the affair, at last met with the surgeon of the vessel, from whom he learnt that before the crew were seized with the disorder, they had been off the north of Ireland, and had had some communication with the inhabitants of those parts.

"But, admitting it should be, by unquestionable evidence, established that the disorder, in some instances, broke out on board ships, to which it could not possibly have been conveyed by personal communication, it will by no means follow that it was not generally propagated in that manner."

CHAPTER V.

ON THE SPREAD OF INFLUENZA BY CONTAGION.

Influenza is contagious, and is chiefly, if not entirely, spread by contagion. Nature of the evidence offered in support of these views.

Enough has been said to prove that some of the highest medical authorities, men of world-wide reputation, teach that influenza is not contagious.

Others admit that influenza is contagious, but teach that contagion plays but a small part in its spread. *Influenza is contagious.*

I do not agree with them. On the contrary I hold, not only that influenza is contagious, but also that it is chiefly, if not entirely, spread by contagion. *And is chiefly, if not entirely, spread by contagion.*

It may be convenient here if I mention the nature and bearing of the facts on which I rely to prove these statements.

(i.) *Cases pointing to contagion.*

In the first place there are a number of cases pointing to the probability that one person derived the disease from another. For example, a man who has been exposed to the contagion of influenza during a visit to an infected town, returns to his native village, which has so far been free from the disease. On his arrival he is taken with it. A day or two later other members of his family are seized, then the people who come to nurse or to visit the sick. In the course of a few days or weeks the disease spreads to many people, in other words, there is an epidemic.

Much of the evidence then consists of cases pointing to contagion.

(ii.) *Isolated cases precede an epidemic.*

In the last chapter I quoted cases which appeared to point to a sudden infection of a large part of the population at the same time, but none of these instances are well authenticated, and I do not believe them to be authentic. It has been afterwards noticed on many occasions that the earliest cases of influenza have for a time escaped recognition. Even when isolated cases have been recognised, little or no importance has been attached to the fact that each case of the disease is centre of infection. Thus, Dr. Frank Clemow, speaking of the early isolated cases, says : * "They may be looked upon, to use Sir Thomas Watson's words, as the first dropping of the thunder shower . . . . The arrival of the great wave of infective material was not until much later, and must be taken as indicated by the occurrence of the disease in large numbers of the population." I believe this teaching to be quite erroneous. I cannot believe in a "wave of infective material." I am content to adopt Watson's metaphor, but in a sense entirely different

---

* Public Health, Vol. II.

to the one in which he used it. The first droppings of a thunder shower point to a coming storm. The first cases of influenza point to an impending epidemic: but they do more; they produce it by contagion.

(iii.) *Influenza spreads along the lines of human intercourse.*

Other facts bear on a geographical consideration of the spread of the disease. It is found that large towns are affected sooner than small ones, and towns sooner than the villages around. That is to say, influenza spreads along the lines of human intercourse.*

(iv.) *Isolated persons, such as prisoners and inmates of asylums and convents, often escape influenza.*

Much important knowledge is derived from the records of asylums, prisons, and nunneries, places more or less cut off from contact with the outer world. In such places it is found that if influenza be not introduced from without, the inmates do not have it. If the disease be introduced it often spreads rapidly, especially under unhygienic conditions.

In the chapters which follow, then, I shall bring evidence to prove—

(1.) That influenza spreads from the sick to the sound by contagion.

---

* Some evidence refers to the possible spread of the disease by parcels, and on this point more information is wanted.

(2.) That isolated cases of influenza precede an epidemic.

(3.) That influenza spreads along the lines of human intercourse.

(4.) That prisoners often escape influenza, although the disease may be raging in the town in which the prison is situated.

CHAPTER VI.

THE SPREAD OF INFLUENZA BY CONTAGION IN 1782 AND 1803.

Importance of the writings of former English physicians. Gray's account of the epidemic of 1792. Cases of contagion mentioned by Dr. Clark, of Newcastle, Dr. Samuel Foart Simmons, Dr. Hamilton, Dr. Haygarth, of Chester. Cases of contagion recorded in the epidemic of 1803 by Mr. Lawrence, of Cirencester, Mr. Harness, of Tavistock, Dr. Fowler, of Salisbury, Mr. Hugo, of Crediton. The death rate in London during the epidemic of 1782.

A very large amount of information bearing on the question of contagion could be collected from the works of those of the old English physicians who have written on influenza. Our own writers have made some of the most exact observations on the disease, and these have received an unusual amount of attention even from Continental authorities. For example, in the official report which he delivered to the Minister of the Interior, on the epidemic which occurred in Paris in December 1889, Professor Proust referred to the works of Huxham and of Munro; and Professor Bäumler, in a

Evidence of contagion collected in former epidemics.

pamphlet he has lately published, attaches great importance to the writings of other English authors.

Dr. Theophilus Thompson's book, "Annals of Influenza" is almost entirely made up of quotations from standard English works on the subject, and scattered through this excellent compilation there are many striking cases pointing to contagion. It seems strange that these have not attracted more attention from modern readers, for the book is still a classic. I believe that this neglect has been to a great extent caused by the fact that Dr. Thompson did not confine his remarks and quotations to matters bearing on the subject of influenza, but laid almost as much stress on other interesting subjects, as, for example, the occurrence of earthquakes, the appearance of comets, and the visitations of insects. The casual reader is apt to become puzzled at the variety of matters dealt with, and important facts bearing directly on influenza are in consequence occasionally overlooked.

I shall quote a few cases from the older authorities, and then pass on to the facts which have been noticed during the epidemic of 1889, 1890, 1891.

Dr. Edward Gray's account of the epidemic of 1782. An excellent account of the epidemic of 1782 was written by Dr. Edward Gray,[*] who had the advantage of receiving information from all parts of England.

The following note on the outbreak of influenza at Newcastle was supplied by Dr. Clark.

---

[*] Medical Communications, Vol. I. London, 1784.

"The disease first made its appearance at Shields, the port of Newcastle, on or about the 26th May, and what is remarkable, before it seized any person in the town, some ships had arrived from London, where the disease was epidemic, whose crews had laboured under the distemper on their passage. And on the 27th and 28th of the same month, a very considerable number of vessels came into the harbour from the river Thames, after a sail of little more than forty-eight hours. The first family (at Newcastle) as far as can be ascertained, was seized on the 28th May, and as the persons who were attacked kept a public shop, it is more than probable they received the infection from some sailors who had arrived from the ships at Shields. This opinion of the disease being introduced into Newcastle by infection, is further confirmed by the following fact, for which I am indebted to Alexander Adams, Esq. The master of a vessel who arrived at Shields in forty-eight hours after he left the river Thames, came to his office on the 28th May, labouring under the distemper. On the 29th, one of the clerks in the office was seized, and, as far as I can learn, was the second person who was attacked with the disease in town."

Similar cases have lately been noticed in England and Ireland. Thus, influenza was introduced into Westport, county Mayo, by a foreign sailor,* and still more recently the origin of an outbreak at Hull has been traced to the crew of an American ship.†

---

* British Medical Journal, 1890, Vol. I., p. 35.
† Private letter from Dr. Daly.

Dr. Simmons' paper.

Samuel Foart Simmons, M.D., F.R.S., in the "London Medical Journal," Vol. IX., year 1788, wrote: "In the late epidemic, as in former diseases of the same kind, many facts occurred tending to corroborate the opinion of its being propagated by contagion."

Here is a case on which he relies in proof of his proposition:

"A lady who came from Suffolk on a visit to a family in London on the 23rd of July, found several persons of the family labouring under a disease. She herself was seized with it on the 30th of July, and on the 1st of August she returned home; but was so ill after she got back into the country that she was confined for several days to her bed. The disease had not then made its appearance in her neighbourhood, but on the fourth day after her return one of her daughters became affected with it, and in the course of about three weeks it went through the rest of her family, which consisted of six persons."

Instances of contagion at Ipswich.

Dr. Hamilton* has given the following account of an outbreak which occurred at Ipswich in 1782.

"A surgeon at Ipswich happened to be in London at the time it raged in it; he left it on the last day of May, and arrived at his house about 11 next morning. 'I left town,' he says, 'the last day in May at night, and was then ill of it. I had none under my care then in it; a few days after, I had several, but none so much debilitated as myself.' From him it spread through all the town."

---

* Memoirs of the Medical Society of London, Vol. II.

Dr. Hamilton* has recorded the following facts: "The first who were seized with it at Norwich (I have it from good authority) were two men lately arrived from London, where it then continued to rage. A sergeant of grenadiers of the 10th regiment of foot, went to London on furlough; the disease then raged in the capital; he returned in a few days to St. Albans, affected, and communicated it to the people in whose house he had his billet. This was the first of its appearance there, and from thence it rapidly spread all over the town."

And at Norwich.

One of the most valuable contributions to the literature on influenza is Dr. Haygarth's Dissertation "Of the manner in which the influenza of 1775 and 1782 spread by contagion in Chester and its Neighbourhood." I have not been fortunate enough to see the original paper, and the following quotation is given on the authority of the late Dr. Theophilus Thompson, who has not stated where or when the Dissertation was published.†

Dr. Haygarth's account of the spread of influenza at Chester.

"As the first patient I had seen in the influenza of 1775 was the landlady of a principal inn, and as I had observed so distinctly that the epidemic of 1782 was brought into Chester by a patient coming from London, I stated this question to my correspondents: 'Could you discover whether the distemper was introduced into your town from any

---

* Op. cit.
† I failed to find the paper at the British Museum, and at the libraries of the Royal College of Physicians, the Royal College of Surgeons, the Medico-Chirurgical and Medical Societies. It is not in the University Library, Cambridge. I have to thank the librarians of these institutions for their help.

place where it had previously attacked the inhabitants?'

"My answers were, '(1.) That the first patient who had the disease in Frodsham was seized with it as he was returning thither from Manchester. (2.) That at Malpas the first patient was the landlady of the inn and her family, a week sooner than any other patient in the town. (3.) That the first person who had the distemper in Middlewich brought it from Liverpool. (4.) That the first person affected with the influenza at Mold had been at Chester a few days before, in a family ill of that distemper. (5.) That a gentleman arrived at Oswestry ill of the influenza, before the inhabitants were attacked. (6.) That at Tarporley the first person seized was a postilion, who had driven a chaise thither from Warrington, where the distemper had previously appeared. (7.) That at Wrexham the first patient came from Chester, and the second from Shrewsbury."

This series of observations appears to me to be of the greatest value.

The epidemic of 1803.

The next cases I quote occurred in the epidemic of 1803. The following were recorded by Mr. Lawrence,* of Cirencester :—

Instances of contagion at Cirencester.

"The first case of well-marked influenza I saw I believe was on the 5th of March; it happened to be a robust and healthy farmer, who the week before, on a journey into Essex, passed twice through London: he lives in a village about five

---

* Medical and Physical Journal, Vol. X., p. 200.

miles from hence, where, at that period, the influenza had not appeared, but was then universal in London."

Mr. Harness,* of Tavistock, made the following observations :— <small>Tavistock.</small>

"Two ladies of this place spent a few days at Exeter and slept at a friend's house where the family had been ill of the influenza (and, indeed, some part of it then laboured under the complaint); one of them was seized as she was returning, and the other two days after; and the whole family where they lodged had the complaint within ten days. About the same time, a person coming from Plymouth Dock, where the influenza was very prevalent, was seized at a friend's house at a different part of the town from the ladies just mentioned. The family of this house likewise soon became infected. These were the first instances of the complaint in this town, but it soon became general."

At Salisbury Dr. Fowler† noticed the following striking facts:— <small>Salisbury.</small>

"One large family in the country, and who had little communication with others, escaped the disease till June (*i.e.*, for two months). They thought they caught it from their music master. Seven persons, who attended in succession a lady

---

* Op. cit., p. 291.
† Op. cit., p. 386.

who had it severely, were attacked with it. Her daughters, who were kept away from her, escaped."

Crediton.
Mr. Hugo[*] gave an excellent account of the outbreak which occurred at Crediton.

"The first case (says Mr. Hugo, of Crediton,) which came under my observation was on the 22nd of March, in the family of a gentleman who resides about three miles west of this town. He had been attending, with his lady, the assizes at Exeter the whole of the preceding week, at which time the influenza was very general there. They came home both ill of the disease. On the next day the servant who returned with them was seized with it, and by the 25th it had been communicated to every other person in the house. Some labourers who resided at an adjoining farm were affected about the same time; but a woman who had been employed at Exeter was the first attacked by it. It appeared very soon afterwards in the town of Crediton; and here also the first case I visited was a gentleman who had been attending at the assize. It spread very rapidly, and in a short time became general in the town and adjoining villages."

Summary of evidence.
From these few quotations it is evident that the epidemics of 1792 and 1803 afforded opportunities for observing some of the facts on which I rely to prove that contagion plays a great part in the spread of influenza. Where careful observations were made it was seen that the first patient in a town or in a rural district had brought the disease

---

[*] Op. cit., p. 291.

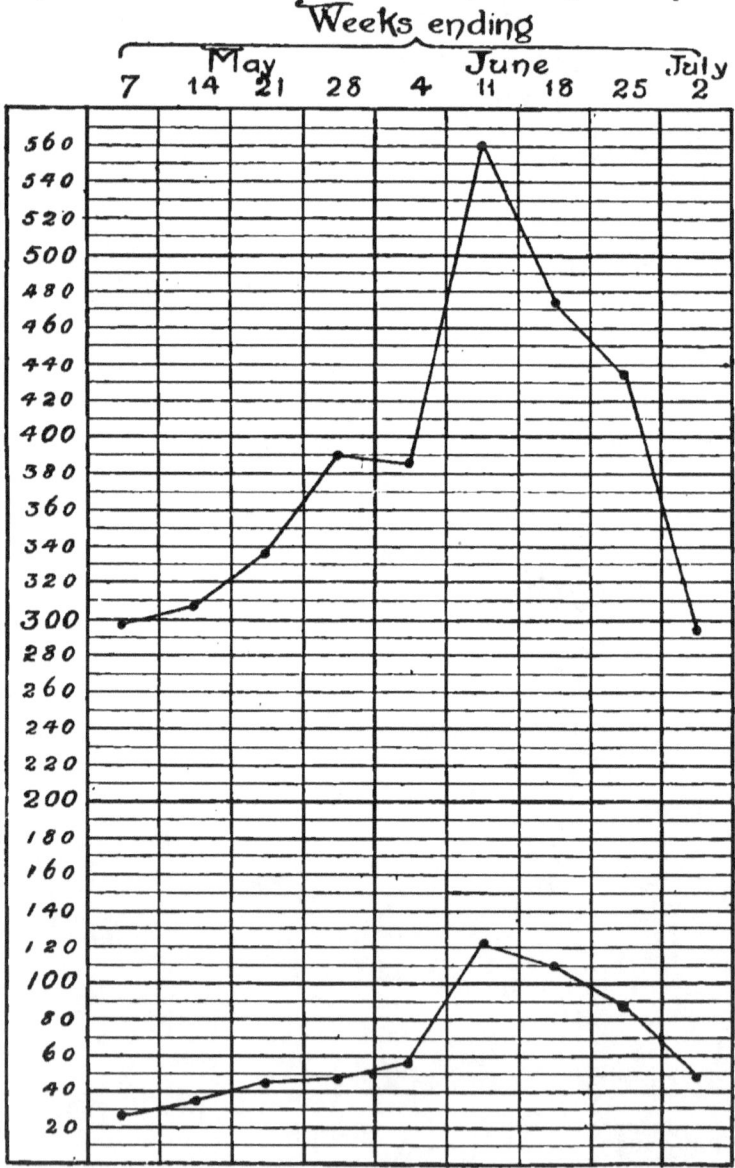

CHART I.

Chart shewing Bill of Mortality and Number of Deaths recorded under the general name of Fever in London in the year 1782 for Weeks ending

with him from some infected place, and that isolated cases preceded a general infection. These facts were commented on rather quaintly by Dr. Hamilton.* "Thus," he wrote, "we find one taken ill to-day, a second to-morrow, and a third perhaps not till several days after. Is not this," he asks, " the usual mode of seizure in all contagious diseases?"

It is interesting to compare the death rate which occurred in London in 1782, during an epidemic of influenza, with that during the visitation of 1889–1890, page 69. The Chart (I.) was prepared from figures given in "An Account of the Epidemic Disease called the Influenza of the year 1782. Collected from the observations of several Physicians in London and in the Country by a Committee of the Fellows of the Royal College of Physicians in London."† This paper was read at the College on June 25th, 1783, and contains some remarks which may be here quoted.

{The death rate in London in 1782.}

"The great increase in the burials after the disease had appeared about three weeks, which is about the time when its effects would most generally be felt, is very striking. At the period when the effects of an acute disease would probably be over, it may be observed that the numbers are again reduced to what they were previous to the appearance of the disease. To strengthen the inference which may be drawn from the foregoing remark on the bills of mortality, notice should be taken of the

{Not a "sudden visitation."}

---

* Memoirs of the Medical Society of London, Vol. II., p. 447.
† Medical Transactions published by the College of Physicians in London, Vol. III. London, 1784.

increase of the numbers of those who are there mentioned as having died under the general head of fever, keeping pace exactly with the periods above pointed out."

The chart shows, then, that the death rate rose as the epidemic became general. In the increase in the number of deaths we see no "sudden visitation," but a gradual rise to the maximum, and a gradual fall, as in the case of other contagious diseases.

CHAPTER VII.

THE EPIDEMIC OF 1889-1890. BOKHARA. ST. PETERSBURGH. BERLIN.

The epidemic of 1889-90. Influenza occurred in Bokhara in May 1889, and the height of the epidemic there was in July. St. Petersburgh was affected in October, and the epidemic reached its height there in November. Berlin was affected in November, and the height of the epidemic occurred there in December. Professor Bäumler's observations on the spread of Influenza by contagion.

Influenza, we have seen, is endemic in some parts of China, and Central Asia has from very early times been looked upon as the source of the disease. The epidemic of 1889-1890 has not been traced so far as China, and most of the writers on it have been satisfied to begin their history of the visitation with an account of the appearance of the disease in Russia.

Dr. Heyfelden, in a paper he contributed to the "Unsere Zeit,"* gave evidence which proves conclusively that influenza was prevalent at Bokhara long before St. Petersburgh was affected.

---

* Unsere Zeit. Leipzig, 1890, p. 185.

Influenza in Bokhara in May 1889.

[*For a full account of the disease as it appeared in that city I must refer the reader to Dr. Heyfelden's paper, but as this has not been published in English I give a brief resumé of its chief points.*]

Dr. Heyfelden believes that the epidemic commenced at Bokhara in May 1889, and was prevalent there from that month until August. Europeans, amongst whom were railway servants and soldiers, were the first people affected, and the disease afterwards spread to the Jews and natives of the city. The characteristics of the disorder were malaise, nausea, shivering, a high temperature, and various nervous symptoms. The fever lasted from one to five days, at the end of which time the temperature often became subnormal. During the attack profuse sweating occurred. The appetite was bad. Great nervous prostration succeeded the febrile stage. There is no reason to doubt, from this description, that influenza was prevalent at Bokhara. A question of great interest is, what was its origin? Was it imported, or did it arise there? To these questions Dr. Heyfelden's paper gives no decided answer. It is true that he does not suggest that it was imported, and he does suggest that influenza is really a severe form of a well-known endemic malady. But the evidence is not conclusive. It seems that a malarial fever is prevalent at Bokhara in the late summer of every year, and when people began to suffer in the way just described some physicians were of opinion that this malarial fever had appeared earlier than usual. This they accounted for by the fact that the ground water was high and the country generally was unusually damp.

Two reasons were given why the people were more susceptible than usual to such an outbreak. In the first place, they had suffered much from cold during the unusually severe winter of 1888–89, because their dwellings were only adapted for summer use. They had consequently spent what little money they had in fuel rather than nourishment, so that they had been underfed and were half starved. The severe fast of Ramadan increased the wretchedness of their physical condition. In the month of May the people were so prostrate that even the youngest and strongest often fainted, or were sick on taking their first bite or their first sip. The second cause which contributed to the wretchedness of their condition was the fact that many of them suffered from the Filuria (*sic*) Medinensis und Buchariensis, and this parasite was more prevalent than usual because there had been a drought during the summer of 1888, and the people had drunk to the bottoms of the wells, to which the ova of the worm naturally gravitate. The fever thus attacked a weakened, bloodless population. It was computed that half the inhabitants suffered, and that of a population of eighty or a hundred thousand from five to seven thousand died. This estimate was perhaps excessive, for some deaths occurred from typhus and typhoid fever, which are always extremely prevalent in a city like Bokhara, where, between high walls and in a small space, the people are crowded together with horses, cattle, sheep, and even camels, and where also little attention is paid to hygiene.

Height of the epidemic there in July.

The height of the epidemic was in July, when whole households, including little children, were simultaneously affected. At one time the servants at the Russian Embassy were nearly all suffering, so that there was no one either to cook the meals, or to perform ordinary every day duties.

St. Petersburgh affected in October.

Dr. Heyfelden has no doubt that the disease he saw at Bokhara was the same as that which afflicted St. Petersburgh in October 1889, and which occurred, as he says, " first in isolated cases then in heaps."* Amongst the first sufferers in St. Petersburgh those who had previously had malarial fever supposed they were suffering from a relapse or recurrence, and they were strengthened in that conviction when it happened that treatment by quinine appeared beneficial. It was soon observed, however, that people who had not had malaria were afflicted in the same way, and after a time it was recognised that influenza was present, and the fact was noticed first by the lay and afterwards by the medical portion of the press. At length there was an epidemic; trade was much interfered with, the performances at the theatres had to be changed at short notice because of the illness of the players; those who lived in common—as scholars and soldiers—were afflicted in greater number than those who lived more private lives.

---

* Im October 1889 traten in St. Petersburg nach einem feuchten, kühlen Sommer, dem ein aussergewöhnlich warmer Herbst gefolgt war, erst vereinzelt, dann häufiger acute Fieberanfälle auf. Unsere Zeit, 1890, p. 188.

CHARTS II. AND III.

Chart shewing the Death Rate per annum per 1000 living in St. Petersburg during the Influenza epidemic in 1889-90 for Weeks ending

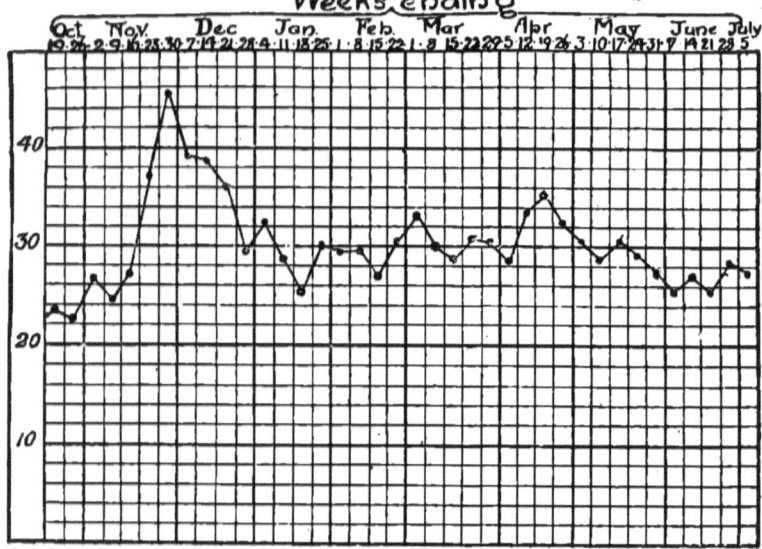

Deathrate per annum per 1000 living in Vienna during the epidemic in 1889-90 for Weeks ending

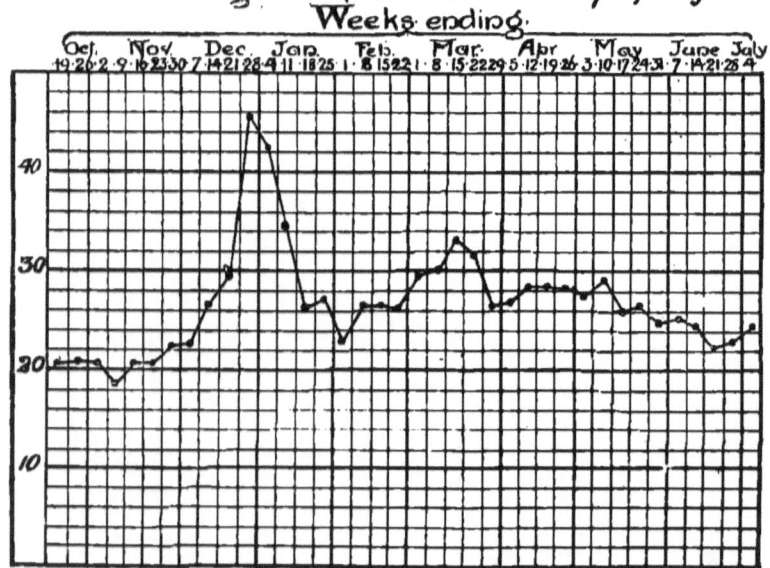

In the beginning of November, Moscow and other Finnish towns were affected by the disease, and news received from the Crimea, and from Tiflis, confirmed the appearance of the disorder in those places.

A physician who left Chabarowka at the end of September, and reached St. Petersburgh at the beginning of December, brought the news that he had recognised some cases of influenza, first at Omsk and afterwards in all the other Siberian post stations which he had to pass between that town and the capital.

At St. Petersburgh the height of the epidemic occurred at the end of November, from which time the number of cases gradually lessened.

Height of the epidemic there in November.

I wish to call attention to the fact that both at Bokhara and at St. Petersburgh isolated cases of influenza preceded the epidemic.

The accompanying chart shows the increase in the death rate which occurred in St. Petersburgh during and after the epidemic. The accompanying Chart (II.) is prepared from the weekly returns given by the Registrar-General, and shows the death rate from the week ending October 19th to the week ending July 5th. A glance at the chart shows—

- (i.) That the highest death rate was not reached suddenly.
- (ii.) That the highest death rate corresponded in point of time with the height of the epidemic.
- (iii.) That after the height of the epidemic the fall in the death rate was slower than the rise which preceded it.

Exactly the same things are to be noticed in the Chart (III.) which is drawn to the same scale, and shows the death rate which occurred at Vienna during the same period.

Before the end of November influenza had reached Berlin, and, according to Professor Leyden, about a third of the inhabitants were affected.

Berlin affected in November.

The Chart (IV.) shows the death rate which occurred in Berlin before, during, and after the epidemic. It corresponds in all respects to charts representing the mortality in the other cities. The highest death rates, it is seen, occurred at Berlin and Vienna during the same week, a month later than in St. Petersburgh, and a week earlier than in Paris.

Height of the epidemic there in December.

Professor Bäumler, in a letter to Dr. Ogle,* gave an account of the epidemic which was present on January 4th, 1890, at Freiburgh, in Baden. Dr. Bäumler noticed that "many patients that were in the wards for other ailments were at once seized when a single case was admitted into the ward."

Professor Bäumler on the contagious nature of influenza.

Professor Bäumler has since published a most valuable pamphlet,† in which he discusses the method of spread of influenza, and comes to the conclusion that contagion plays the chief part. The following case, which he quotes from the "Correspondenzblatt," is of special interest, but the whole paper will repay careful study :—

"Influenza was conveyed to the hotel on the Feldberg, in the Black Forest, by the host, who

---

* Lancet, 1890, Vol. I., p. 103.
† Ueber die Influenza von 1889 und 1890. Wiesbaden, 1890.

CHARTS IV. AND V.

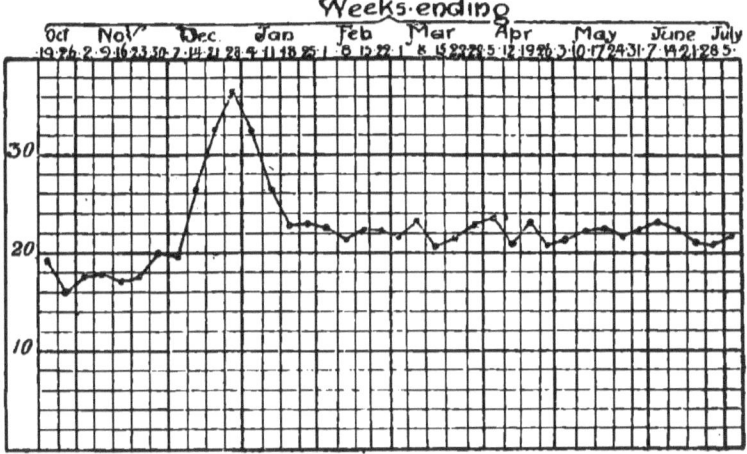

Death Rate per annum per 1000 living in Berlin during the Influenza epidemic in 1889-90 for Weeks ending

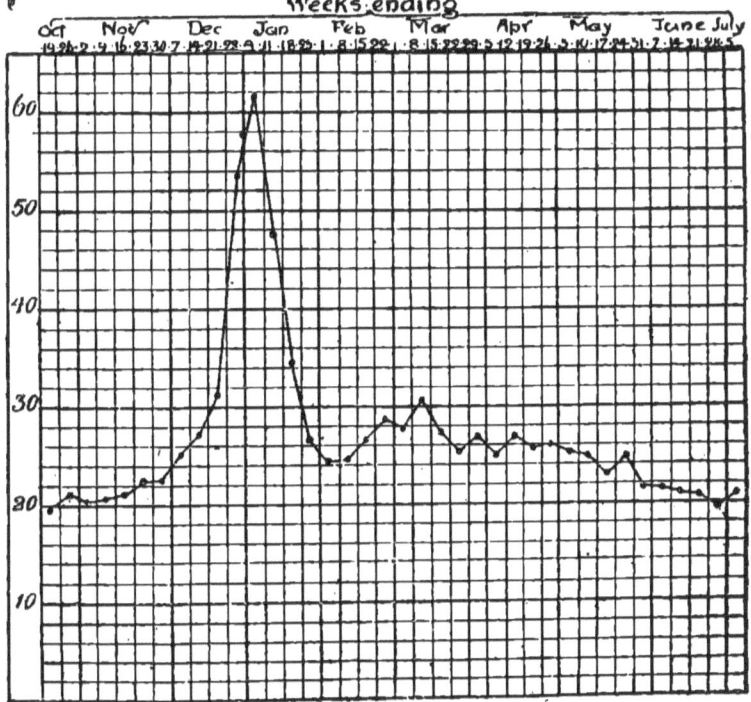

Death Rate per annum per 1000 living in Paris during the Influenza epidemic in 1889-90 for Weeks ending

had contracted it at Freiburgh. The next victim was his sister, who nursed him, then the maid who nursed her, while the men servants, who were much in the open air, escaped, as did also the other occupants of the house who were purposely kept away from the sick room."

This case not only points to the contagious nature of influenza, but also shows the importance of practically recognising the fact by adopting precautionary measures against its spread. The occupants of the hotel avoided the chances of infection, and the disease did not affect them.

CHAPTER VIII.

THE EPIDEMIC OF 1889-1890. FRANCE.

Cases of Influenza were recognised in Paris in November, the height of the epidemic occurred there in December. Influenza was imported to Montpellier from Paris, and isolated cases preceded the epidemic. To Frontignan it was imported from Paris. To Montbéliard it was imported from Neufchatel and Soleure. To Vergèze it was imported from Lunel. Dr. Antony and M. Barth noticed that influenza spread by contagion in hospitals. On board the steamer "S. Germain" it spread rapidly after it had been introduced by a passenger. Influenza spread rapidly on the training ship "Bretagne" after it had been introduced by an officer who was suffering from it. Two other training vessels which were moored near escaped.

The course of the epidemic of influenza which afflicted France in 1889 has been carefully studied by Professor Grasset, of Montpellier, and the results of his researches are to be found in his excellent book, "Leçons sur la Grippe de l'hiver, 1889-1890."[*] I have freely availed myself of the

---

[*] Published at Montpellier, 1890.

valuable information given in this brilliant and carefully written monograph, which I commend to the notice of those who are interested in the question.

Influenza in Paris was first recognised in the shops of the Louvre. Before November 25th, the average number of absentees from 3,000 *employés* was 120. From this date the numbers of those absent rapidly increased, and rose to 515 and 670 on the 8th, 9th, and 10th of December. On December 5th the disease affected the workers at the bureau central des Postes et Télégraphes. On the 10th the epidemic was general. *Influenza in Paris in November.* *Height of the epidemic there in December.*

Thus in Paris there were localised outbreaks before there was any general infection of the general population.

The chart showing the increase of death rate in Paris during the visitation was, like the others, prepared from the weekly statistics given by the Registrar-General. (Chart V., page 52.)

The origin of the outbreak at Montpellier was recorded by Professor Grasset.

"On the 8th of December one of my colleagues returned from Paris, where he had been passing a few days. During his stay in the capital he had gone to make some purchases at the Louvre. On the 9th he was seized with pain in the back and headache, and the disease declared itself. On the following day there were other isolated cases. On the 15th there was a general epidemic." *Outbreak at Montpellier, imported from Paris.* *Isolated cases preceded the epidemic.*

To Frontignan also influenza was imported from Paris, and afterwards spread rapidly.

Outbreak at Frontignan, imported from Paris.

The account given by Dr. Bordone is as follows:—

M. A. went to Frontignan from Paris on December 15th. He was taken ill on the journey and had influenza. On the 17th he dined at his own house, and the party consisted of ten people besides himself. On the 19th five of these people were seized with influenza. These were the first cases in Frontignan. On the 18th M. A. went to his office. On the 21st his *employé* had influenza. This man lived at Vic, a village some kilometres from Frontignan. He was the first sufferer from influenza at Vic. Five days after his seizure the patient's mother, who lived with him, had symptoms of the disease.

From the 23rd of December the disease spread rapidly both at Frontignan and in Vic.

Montbéliard, imported from infected neighbouring towns.

Dr. Tueffart, of Montbéliard, made some valuable observations on the introduction and spread of influenza at Montbéliard. These were communicated to Professor Bouchard, and by him brought to the notice of the Paris Academy of Medicine.*

Influenza made its first appearance in Montbéliard on December 13th, 1889. Before that date the disease was prevalent in the neighbouring towns (Neufchatel, Locle, Chaux-de-Fonds, Bienne and Berne). On the 6th of December an inhabitant of Montbéliard remained for a great part of the day in a hospital containing patients suffering from influenza. He returned to Montbéliard and was seized with the disease on the 13th. On the 17th

---

* Bull. Acad. Med., Paris, 1890, 2 s. xxiii.

his two daughters were similarly affected. On the 19th his son began to suffer.

This young man had a friend with whom he was brought into contact daily. On the 20th the friend had the disease. On the 21st the father of the latter took it. On the 23rd the brother-in-law of the last-named was seized.

On the same day the wife of the man who first failed with the disease was attacked, and at the same time three young people, friends or relations of the latter. Thus in ten days, from one source, the disease apparently spread to eleven people.

While this was happening influenza was being imported into the town by other people.

On the 21st it broke out at the house of a merchant who had lately returned from an infected house at Neufchatel.

On the 22nd the disease was brought from Soleure by another tradesman.

After reading this case M. Bouchard insisted strongly on the importance of the fact the influenza first manifested itself in isolated centres, and that there was not a sudden visitation.

He said: "One does not find in the same degree the sudden and universal invasion which rightly or wrongly was said to have characterised previous epidemics, characteristics which made me attribute influenza to meteorological causes, and led me to say at our meeting of December 17th last that influenza did not seem to be contagious nor even infectious."

Professor Trueffart's observations did not convince M. Bouchard that influenza was contagious.

From his own experiments, however, he came to the conclusion that the complications of influenza, and particularly pneumonias, were contagious; and this he thought explained the fact which he had recognised that at the decline of an epidemic, when influenza properly so called disappears, cases of pneumonia are often very plentiful.

Lunel. Instances of spread by contagion. An account of the outbreak of influenza at Lunel was recorded by Dr. Védel, and published by Professor Grasset.* It is as follows:—

On December 22nd, a young girl of 15, an inhabitant of Lunel, was seized with the epidemic; this, if not the first, was one of the first cases which occurred in the town. Between the 22nd and the 26th the girl's two brothers and her mother were taken in their turn. The father only was spared, or at most was but lightly affected. On the 24th four of the five members of the household were in bed, and a friend who lived at Vergèze (8 kilometres from Lunel) came to visit the family, and stayed with them for about two hours. On his return to his own house on the same evening he was immediately and sharply attacked. On the following day, the 28th, his wife, and afterwards his two daughters, were seized. Then his sister-in-law, who was living in the village, and who had gone to nurse them, was taken. These were the first cases in Vergèze; from the time they were seized the disease spread.

---

* Op. cit.

From observations he made at the hospital of Val de Grâce, Dr. Antony* came to the conclusion that influenza spread by contagion amongst the patients, and M. Barth† noticed the same thing at l'hôpital Broussais. *Spread by contagion in hospitals.*

At the same meeting of the Academy of Medicine at which M. Bouchard related the history of the outbreak at Montbéliard, Dr. Proust said that he was convinced that influenza was contagious, and in proof of the fact he recorded the following case‡ on the authority of Dr. d'Hoste. *Spread by contagion in the packet S. Germain.*

On December 2nd, 1889, the packet "Saint-Germain" sailed for Vera Cruz from S. Nazaire with a clean bill of health. She put into Pauillac on the 3rd and 4th, and to Santander on the 5th. At the latter place she took on board a first-class passenger who had come from Madrid, where influenza was prevalent. This passenger was seized with influenza on the 6th, the day after his arrival on board. The sanitary state on the ship was good, and no case of influenza had previously occurred. It was not till the 10th, that is to say, four days afterwards, that Dr. d'Hoste was taken with it. He was the first person attacked after the passenger who came from Madrid. The next case appeared on the 12th. Then the malady spread, the number of the sick rapidly increased, and 154

---

* Bulletins et Memoirs de la Société Médicale des Hôpitaux de Paris. Ser. iii., Tome vii., 1890, p. 93.

† Op. cit., p. 97.

‡ Bulletin de l'Académie de Médecine. Ser. 3, Tome xxiii., p. 170.

passengers were seized out of a total of 436, as well as 47 of the crew.

Dr. Danguy des Déserts, the chief medical officer at the training vessel "Bretagne," which was stationed at Brest, has published a history of the outbreak of influenza in that vessel. It is as follows :—

Spread at Brest.

The crew of the "Bretagne" consisted of 850 men. On December 11th, an officer who lived ashore, received two large parcels from Paris, and unpacked them himself: three days later he developed influenza. On the next day and the day after, his wife and his three servants were seized with the malady. These five cases were undoubtedly amongst the first cases seen at Brest.

Spread by contagion in the "Bretagne."

On the 14th, the officer went on board the "Bretagne," and remained on board for 48 hours. On the 15th, the first case of influenza was seen on the vessel. From the 17th, there was an epidemic of influenza amongst the crew, and from 25 to 40 new cases occurred daily. Some of the officers and non-commissioned officers were allowed to go to their own homes for treatment. In every case the disease spread to the families of those officers and non-commissioned officers. At the time of the epidemic on the "Bretagne," there were two other training vessels moored near her, the "Borda," and the "Austerlitz." No case of influenza occurred on either of those vessels.

This case is interesting in several ways. In the first place, there is the evidence in favour of the importation of influenza by goods. In the second place, there is the strongest evidence that the disease was taken on board the "Bretagne" by the officer, and that it spread by contagion.

We have seen that it is taught by some physicians that the disease is spread by a general aërial contamination. On this hypothesis how is it possible to explain how it was that the air around the "Borda" and the "Austerlitz" was not contaminated, and that the air around the "Bretagne" was charged with the infective material of influenza? Surely it is more reasonable to suppose that the crew of the "Bretagne" was affected because the infective matter was conveyed to the ship by the officer who had the disease, and that the other ships escaped because the crews were not exposed to contagion.

From France, then, we have on the one hand philosophical speculations by M. Colin and others in favour of an almost instantaneous unexplained aërial contamination, and, on the other hand, we have carefully observed facts in favour of the spread of influenza by contagion. I prefer to accept the facts.

CHAPTER IX.

THE SPREAD OF INFLUENZA IN ENGLAND IN 1889, 1890, 1891.

It is impossible to determine the exact date on which influenza first appeared in England in 1889. Dr. Buchanan says there were cases in October. Dr. Shelly, of Hertford, recognised cases of the "gastric type" in November. During the second week in December there were cases at Westbourne Grove, these were followed by a local epidemic, which reached its height on December 17th. Hammersmith was affected during the third week in December. During the first week in January there was an outbreak at the General Post Office and at Dr. Barnardo's Homes. The height of the epidemic amongst the casualty patients of St. George's Hospital occurred during the second week of January. At Broadwood's pianoforte factory there was a severe epidemic. It was not a "sudden visitation," but increased steadily in severity daily, from December 27th to January 15th, and from that date there was a steady decrease in the numbers of those affected.

Colchester, Canterbury, Chelmsford, Cardiff, and Oxford were affected during the first week of January. Isolated cases existed before there was an epidemic. Many persons employed at the University Press at Oxford were seized. There was no "sudden visitation," but the course of the epidemic resembled that which occurred at Broadwood's factory. During the second week in January isolated cases occurred at Bristol, Birmingham, and Sheffield; epidemics followed.

Many cases were brought into Liverpool from London, but no epidemic followed. At Nottingham there were cases during the second week in January 1890, *the height of the epidemic occurred during the third week in February. At Manchester there were cases as early as the first week in November* 1889, *the height of the epidemic was reached during the third week in February* 1890. *Inverness and Stourbridge were affected during the first week in February* 1890. *influenza was very prevalent at Trowbridge, Wimborne (Dorset), St. Ives (Cornwall), and in the rural parts of Derbyshire about the middle of February. The disease was carried to Churchingford (Devon) from Paris in December* 1889, *it then spread only in the family of the patient first affected, and to the doctor. But an epidemic occurred in February after influenza was introduced from neighbouring towns. As late as March the disease was spreading in Wales.*

During the summer and autumn of 1890 *a few scattered cases occurred in England, and some were seen during the winter of* 1890–1891.

In March 1891, *influenza was imported into Hull from New York, it spread rapidly in Hull. An epidemic at Sheffield followed, and the disease afterwards spread to London.*

Cases in which influenza was spread by contagion have been noticed by Dr. Frank Clemow and Dr. Shelly, of Hertford. At two outbreaks which occurred at Haileybury College the visitations were not sudden, but the number of those affected increased gradually, as in the case of other contagious disorders.

We have already seen that in foreign cities it was found difficult to ascertain the date of the first appearance of influenza. At St. Petersburgh dates as far apart as the beginning and the end of October were mentioned in this connexion. In Paris, too, one great physician spoke of the disease as a sudden visitation coming from no one knew where, but more careful inquiry showed that the visitation was not a sudden one.

In a city like London the difficulty of finding the date of early cases of a disease is even greater than in these—the largest—continental cities, and it is impossible for a private individual to obtain all the necessary information. On this matter no doubt the report of the Local Government Board will throw much light, but it is practically certain that many of the earlier cases of influenza which were treated privately, will never be reported in any public records.

Dr. Buchanan,* at a meeting of the Medical Society on February 17th, 1890, said that mild cases of influenza had occurred in 1889 as early as October, both in London and on the low hills above the Humber.

It is certain that in England, as elsewhere, early cases of influenza often passed unrecognised. Practitioners who had not previously seen the "gastric type" of the disease could hardly be expected to recognise it at first, for the symptoms resembled those which are usually ascribed to catarrh of the bile-ducts, much more than those which had been popularly associated with epidemic influenza. Dr.

---

* Lancet, Vol. I., p. 406., 1890.

Shelly, of Hertford, has kindly sent me an account of such an outbreak which occurred in his practice in November 1889. Writing of influenza he says :—

"The first case which I saw and recognised occurred early in November 1889 at Haileybury, and it was of the gastric type and was complicated with catarrhal jaundice; a group of similar cases followed, extending up to December 18th, five having jaundice. About this time I saw in private practice, in about a fortnight, between 20 and 30 similar cases in persons of all ages, all developing more or less jaundice, and you will recollect that at this time the daily press recorded several 'epidemics of jaundice' in different parts of the country, whole schools having to be temporarily closed on this account in some cases. From the invasion, progress, and subsequent history of my cases I have little doubt that they were really influenza. My first case of the disease exhibiting the rheumatoid neuralgic symptoms was noted in a Haileybury boy on December 6, 1889, and no other case of this type occurred before the school broke up about December 20th, 1889."

Outbreak in Westbourne Grove during 2nd week of December.

In the middle of December 1889 there was nothing like a general epidemic throughout London, but as early as December 10th there was a localised outbreak of the disease amongst the *employés* in a large shop in Westbourne Grove.* A short paragraph in one of the lay papers was the first intimation I had of this, and it led me to ask Mr. Watson, who attended the patients, to give me some account of what had happened. This he kindly did, and a summary of the chief facts were given in

* Lancet, January 4, 1890.

"The Lancet," together with details of the symptoms in some cases which had come under my own notice.

Mr. Watson saw the first cases on December 9th, and the number of those seen daily rapidly increased till December 17th, from which date they as rapidly declined. The total number of cases seen among the *personnel* of this one establishment was one hundred and seventy. By the end of December there were only one or two cases of the disease there. These patients I saw with Dr. Watson, and was able without hesitation to confirm his diagnosis.

At the end of December, although newspaper reporters and journalists were very active in getting news of the disease, it could not be found that there was anything like a general epidemic.* It is only doing justice to the lay press to say that we are

---

* An article called "The Epidemic. By a Medical Correspondent," was published in The St. James's Gazette on December 30th, 1889. It contained the following passage :—

"It may be considered certain that the epidemic disease now raging on the Continent has not yet reached London as such. Careful inquiries made yesterday at some of the great London hospitals, at public institutions such has Scotland Yard, and among medical men (both consulting physicians and general practitioners) in different parts of London, have elicited the same answer everywhere. One or two isolated cases, in which the symptoms strongly resembled those described on the Continent, have occurred in most quarters."

This article is of interest from several points of view. It appeared some weeks after the occurrence of the outbreak of influenza at Westbourne Park, yet the writer, who had evidently taken considerable trouble to get information from many authoritative sources, had not heard of this localised epidemic. These facts show the difficulty attending such an inquiry in a city like London. A point of still greater interest, is the positive evidence that "one or two isolated cases, in which the symptoms strongly resembled those described on the Continent," had occurred "in most quarters." Thus, in London as elsewhere, there were many isolated cases before there was an epidemic.

indebted to it for much valuable information on the spread of the disease, and that in England, as in France and Russia, it gave the first information. Having acknowledged this obligation, I think I am justified in adding that I think it is a matter for sincere regret that the lay journals have admitted into their ordinary news columns, advertisements of quack and other medicines written by laymen, and, I regret to add, by members of the medical profession.

The outbreak of influenza at Westbourne Park was quickly followed by an epidemic in that neighbourhood and at Bayswater. At Hammersmith there was an epidemic amongst the postmen before Christmas. *Hammersmith, 3rd week in December.*

Early in January there was an outbreak at the General Post Office, and on the 3rd the number of absentees in the postal service in London had more than doubled. At the head office alone there were two hundred extra cases of illness. *General Post Office officials, 1st week in January.*

The disease made great havoc at Dr. Barnardo's homes, where two hundred and fifty patients were ill at the same time. At the Stepney Home alone, of three hundred and forty-five inmates, one hundred and forty-five were ill.

At Limehouse policemen were much affected. The epidemic was so prevalent in the Docks, and so many Lascars were laid up, that one of the Peninsular and Oriental vessels was fitted up as a hospital for them.

Railway servants suffered severely. On January 5th ten per cent. of the clerical staff at Euston were affected. On the London, Brighton, and South Coast Railway the sick rate was raised fifty per cent.

Epidemic general, 1st week in January.

The epidemic indeed was general, and numerous cases were treated in the hospitals early in January.

The Charts (VI., VII.) show the number of patients treated in the casualty department at St. George's Hospital from December 30th to January 30th. These charts were prepared from the report which I made at the request of the Local Government Board. It will be noticed that the greatest number of men treated on any one day was on January 6th, which was a Monday, and that there were no men treated on Sunday, January 5th. In order, therefore, to get an accurate idea of the gradual rise in the numbers affected, it is necessary to take this fact into account. I have, however, preferred to let the chart accurately represent the facts as observed. In order to get a true idea of the course of the epidemic it is necessary to distribute the number seen on the 6th and give to the 5th (Sunday) a fair share. If this be done, it will be seen that the height of the epidemic affecting the St. George's casualty patients occurred on January 8th, both amongst men and women. The course of the epidemic is for January 5th and 6th represented fairly by the dotted line.

Casualty cases at St. George's Hospital.

Highest death rate at height of epidemic.

A chart (VIII.) is given showing the death rate in London, according to the weekly bill issued by the Registrar-General. It will be noticed that the rate was higher in December than in November,

CHARTS VI. AND VII.

Charts showing the number of casualty out-patients treated for influenza at St. George's Hospital from December 30th, 1889, to January 30th, 1890.

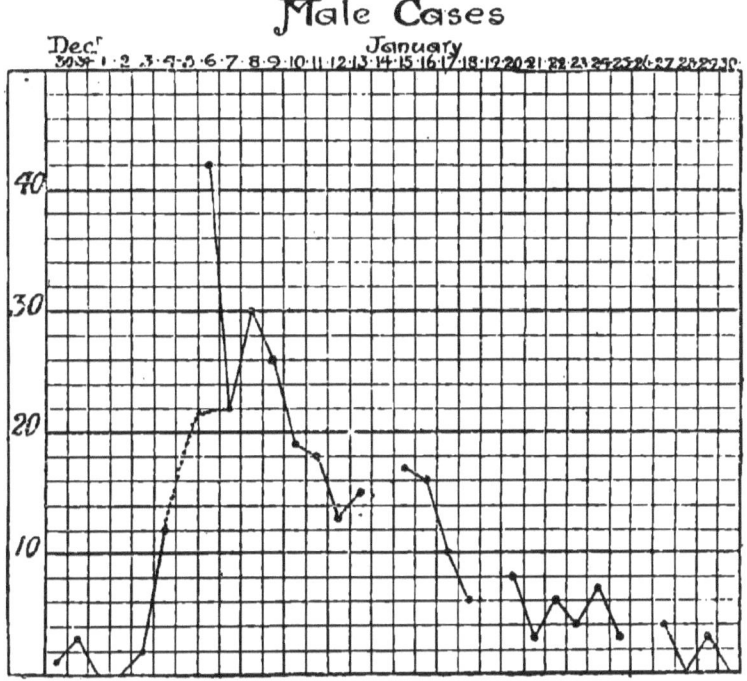

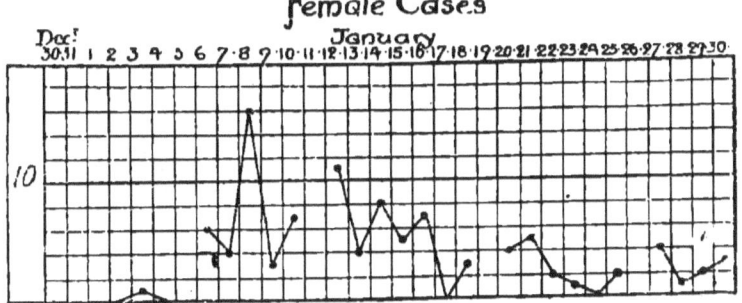

CHARTS VIII. AND IX.

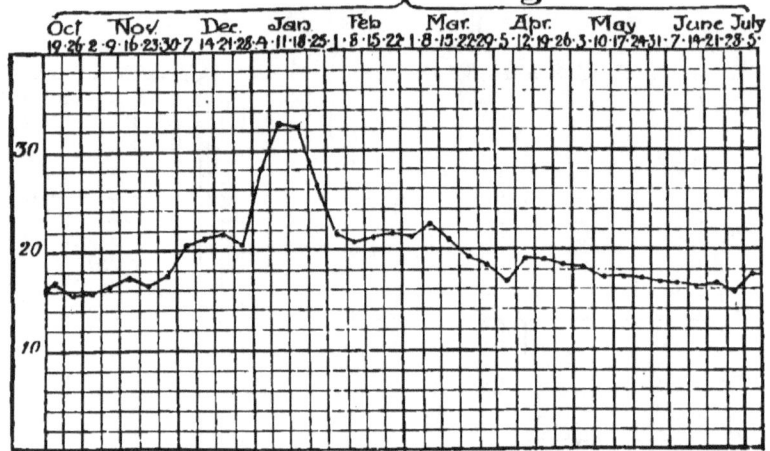

Chart shewing the Death Rate per annum per 1000 living in London during the Influenza epidemic in 1889-90 for Weeks ending

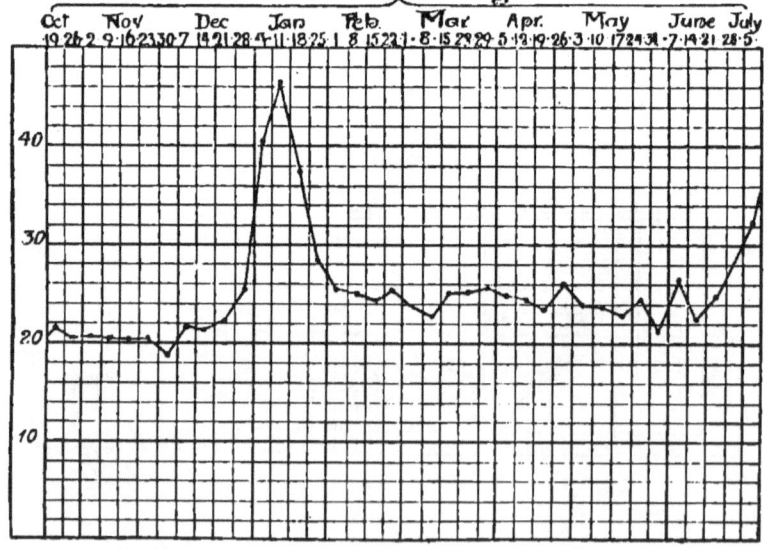

Death Rate per annum per 1000 living in New York during the epidemic in 1889-90 for Weeks ending

and that there was a very considerable rise in January. The highest rate occurred in the week which ended on January 11th. This was the week in which the influenza epidemic reached its height in London. The rise in the death rate, like the progress of the epidemic, was gradual, and there was not a "sudden visitation." A chart (IX.) is given showing the death rate in New York during the same period.

<small>No "sudden visitation."</small>

An outbreak of influenza which occurred at the pianoforte factory of Messrs. Broadwood was investigated by Dr. Sheridan Delépine, with that care and accuracy which characterises all his work. The results of the inquiry were published in the "Practitioner."*

<small>Outbreak of influenza at Broadwood's pianoforte factory.</small>

Broadwood's factory afforded excellent opportunities for such an investigation for several reasons. The sanitary condition of the place was good; accurate records were always kept of the amount of sickness which occur among the men; and the number employed (460) was large enough to make the statistics obtained valuable. Another point was that the workmen were paid by the piece. It was found that people engaged at regular weekly salaries were unusually liable to contract influenza, but no such predisposing cause existed at Broadwood's.

The chart (X.) is prepared from figures given by Dr. Delépine, and shows the daily number of absentees from the factory from December 27th, 1889, to February 4th, 1890. "It might be said,"

---

* Vol. XLIV., No. 4. " Is Influenza a contagious or a miasmatic disease."

Dr. Delépine remarks, "that there is no proof that all the absentees were suffering from influenza. This, I think, is not material. Under normal circumstances the number of men absent through illness is seldom above six during the week at this time of the year. Through an influence of some kind, during the month of January this number rose gradually to 5, 6, 10, 13, 14, 15, 22, 25, 29, 34, 38, till it reached 40 in one day. Besides these men, a few suffered as usual from bronchitis, asthma, consumption, renal disease, rheumatism, erysipelas, &c. During the same month, as is well known, influenza was spreading through London; it is therefore natural to connect the excess of illness in the factory with the epidemic prevailing at the time."

Not a "sudden visitation." A study of the chart shows that there was no "sudden visitation" amongst Messrs. Broadwood's *employés*, but that the numbers gradually increased till the maximum was reached, and then gradually declined.

I must refer the reader to Dr. Delépine's original paper for many interesting details, and for some diagrams and charts, which will be found to be most instructive.

Colchester, isolated cases first, height of epidemic in January. At least as early as January 3rd there were cases in the camp at Colchester, and on the 5th there were 100 new ones. At the same time there were a few sufferers in the town. An epidemic followed. On the 20th of the same month it was declining.

Canterbury, 1st week in January. Early in January Canterbury was affected. The disease broke out in the barracks and in St. Mary's Jesuit College, and in both places the spread was

CHART X.

Chart shewing the total number of persons absent daily from Messrs Broadwood and Sons' Pianoforte Manufactory from Dec. 27th 1889 to Feb. 4th 1890 ill with the Influenza

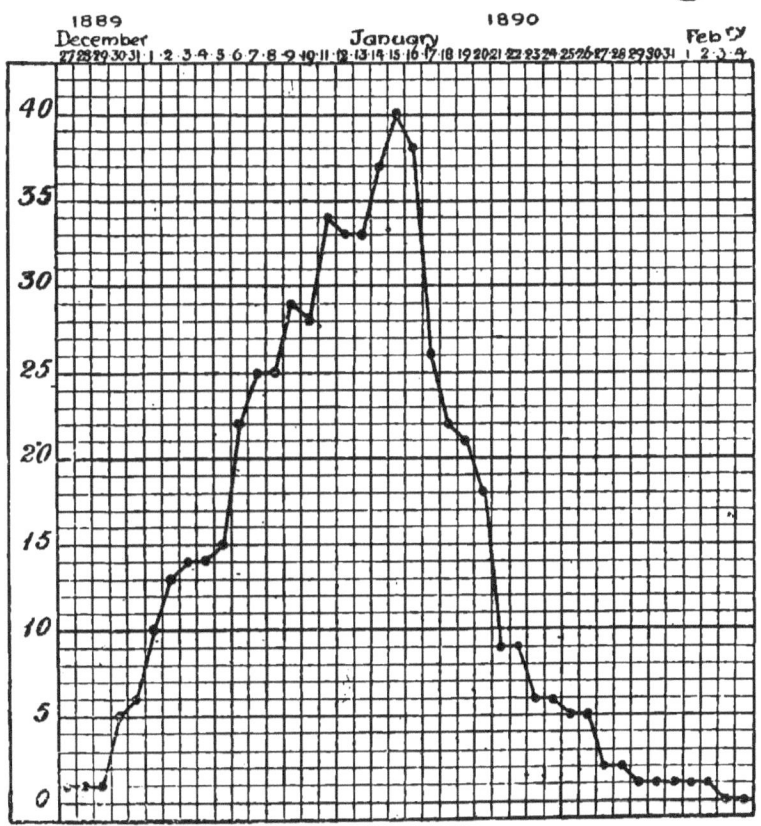

rapid. On January 5th there were 100 new cases among the soldiers, and at the college nearly all the servitors and several of the Fathers were affected.

During the first week of January there were isolated cases of influenza at Chelmsford. The epidemic soon became general. During the third week of January the disease, according to a local paper, was "spreading to all the centres of population which make the county town their head."

Chelmsford, 1st week in January, isolated cases first.

Early in January there were cases at Oxford, but not in the surrounding country.*

Oxford, 1st week in January, surrounding country later.

There was a great outbreak amongst the *employés* of the University Press at Oxford. Mr. Horace Hart, the Controller, published some valuable statistics on the subject. In a letter to "The Times,"† he said:—

"The number of persons employed here at the beginning of January was 562; and out of these, about four are usually absent daily through sickness. But from January 6 onwards, a very different account has to be given. On that day the absences rose to 29; on the following Saturday they were 67; on Monday, the 13th, there was a slight decrease (owing to some of the persons affected coming back too soon); and on Saturday, the 18th, the epidemic reached its highest point, 70 persons being away ill. On Monday, the 20th, we started with 54 absentees, but from that time the number has steadily decreased, until we are now almost back again to the normal number. This state of affairs is shown by the following table:—

Outbreak at the University Press.

---

* British Medical Journal, 1890, Vol. I. p. 148.
† The Times, February 5, 1890.

72 SPREAD OF INFLUENZA IN ENGLAND, 1889–1891.

NUMBER OF PERSONS ABSENT BECAUSE OF ILLNESS from Monday January 6, to Saturday, February 1, 1890, both inclusive.

Date.	Men.	Boys.	Women and Girls.	Total.
First Week.				
January 6	12	16	1	29
,, 7	18	22	1	41
,, 8	20	27	3	50
,, 9	25	25	4	54
,, 10	27	30	4	61
,, 11	31	32	4	67
Second Week.				
,, 13	28	28	2	58
,, 14	30	21	3	54
,, 15	34	27	5	66
,, 16	36	22	8	66
,, 17	33	26	8	67
,, 18	38	24	8	70
Third Week.				
,, 20	31	22	1	54
,, 21	26	21	3	50
,, 22	24	16	5	45
,, 23	28	17	4	49
,, 24	27	15	5	47
,, 25	26	15	4	45
Fourth Week.				
,, 27	15	8	5	28
,, 28	16	11	3	30
,, 29	14	7	1	22
,, 30	14	8	1	23
,, 31	14	6	2	22
February 1	13	2	2	17

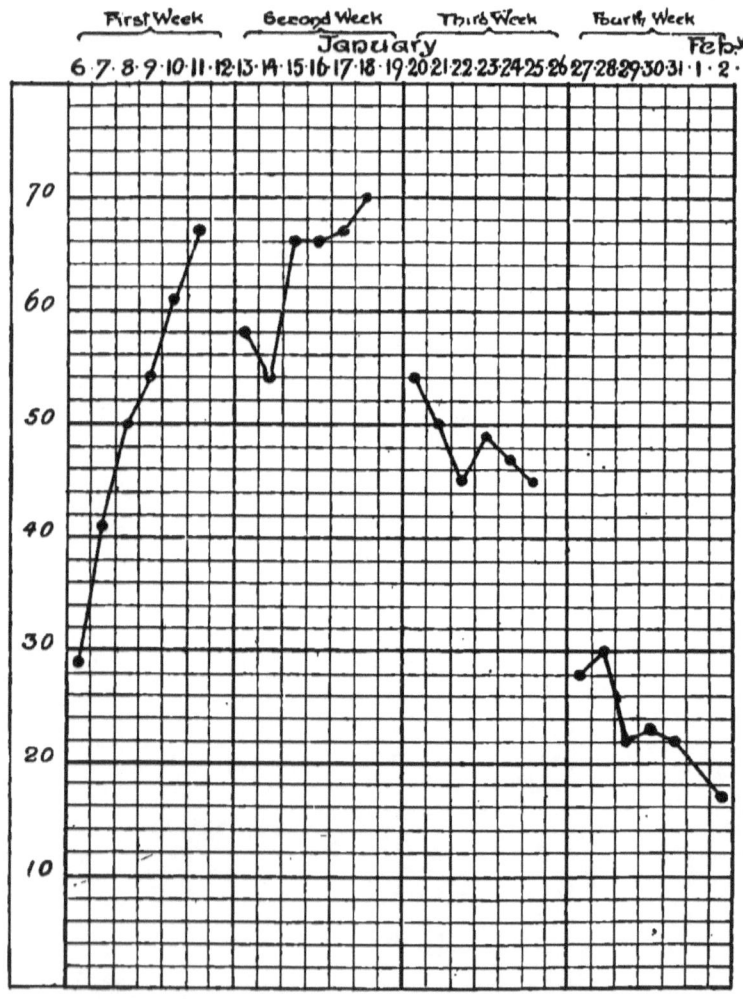

The total number of persons affected during the four weeks was as below:—Men, 108; boys, 103; women and girls, 25; total, 236."

From these statistics a chart (XI.) has been constructed which shows the number of absentees from day to day. It will be noticed that the occurrence of Sunday forms a disturbing element, which prevents the continuity of the most important line in the chart. The point which Mr. Hart's figures teach, and which the diagram clearly shows, is that the outbreak was not a "sudden visitation" but that the disease spread gradually, and, broadly speaking, there was a steady definite rise to the maximum, and then an equally steady fall in the number of those affected. *[margin: Gradual increase in the number affected. Not a "sudden visitation."]*

At Northampton* influenza was prevalent on January 14th.

From Bristol, Dr. E. Markham Skerritt† reported that isolated cases occurred before there was a general epidemic. *[margin: Bristol, isolated cases first.]*

Some of the earliest, if not the earliest, cases of influenza which occurred at Newcastle-on-Tyne‡ were those of sailors from Gothenburg, and afterwards cases occurred amongst those who worked on the quay, and the railway and post office clerks. The epidemic came later. *[margin: Newcastle-on-Tyne, isolated cases first.]*

From Birmingham it was reported, on January 14th, that there were a few scattered cases. The epidemic occurred later. *[margin: Birmingham, isolated cases first.]*

---

* British Medical Journal, 1890, Vol. I. p. 148.
† Ibid., 1890, Vol. I., p. 96.
‡ Idid., 1890, Vol. I., p. 148.

Cardiff, isolated cases first.

The same thing happened at Cardiff. Dr. A. Sheen* wrote, that "after a few mild scattered cases, influenza became widely prevalent about January 8th."

Dr. Taylor,* of Cardiff, reported that "influenza commenced in his practice on January 4th, in the persons of the post office *employés*, several of whom were seen on that day. On the 5th, 6th, and 7th, many more cases were attended to, and during the week ending January 11th, above 50 post and telegraphic *employés* were invalided; several railway men were also laid up."

Liverpool, isolated cases 1st week in January.

At Liverpool isolated cases occurred before there was a general infection. "Cases of the Epidemic Influenza . . . . have also appeared, though these have been isolated rather than epidemic."† The first case occurred about January 4th, in the house of a merchant who was largely engaged in correspondence with France.‡ On January 6th the disease broke out at an emigrant's lodging-house. There were a few other isolated cases about the same time, but Liverpool suffered much less than many other towns.

Dr. J. Stopford Taylor,§ the Medical Officer of Health for the city, has published this report:—

"Liverpool was fortunate in escaping an epidemic outbreak, for many cases were imported from London, and there was no diffusion of the disease, though, doubtless, there was a morbific influence

---

* British Medical Journal, 1890, Vol. I., p. 146.
† Lancet, 1890, Vol. I., p. 211.
‡ British Medical Journal, 1890, Vol. I., p. 96.
§ Report of the Health of Liverpool during the year 1890, by J. Stopford Taylor, M.D. Liverpool, 1891.

in operation, which rendered diseases of the lungs more fatal; yet there was not that contagiousness about it which marked the disease in London and other large cities, only eight deaths being certified as directly due to the disease. Why Liverpool, with its dense population, should have escaped so lightly and other cities suffered so severely is not easy to determine, for in previous visitations we have suffered severely, and its effects were felt in a long-continued high death rate."

In the middle of January there were cases at Weymouth but none at Dorchester, and Dr. William Lush wrote that "In a certain manor house some five miles distant (from Dorchester), a case having been imported from London, out of a family of eight or nine others only one escaped." *[Imported to Dorchester from London.]*

At Sheffield, on January 14th, there were three cases, "all of which have been," it was said, "imported from a distance."* Afterwards there was an epidemic. *[Sheffield, imported from a distance.]*

Dr. Theodore Thompson,† late Medical Officer of Health for Sheffield, has published a table which shows the rise which took place in the death rate during the epidemic, and the number of deaths which occurred from influenza, pneumonia, and bronchitis. Dr. Thompson's table commences with the week ending February 8th, that is to say, more than a fortnight after the appearance of the first cases; and it is to be noticed that the height of the

---

* Letter from Mr. Arthur Jackson, British Medical Journal, 1890, Vol. I., p. 148.
† Public Health, Vol. III., p. 421.

epidemic, as judged by the death rate, occurred about the middle of March.

				Week of Epidemic.	Death Rate per 1,000 per Annum.	Deaths from		
						Influenza.	Pneumonia.	Bronchitis.
Week ending February	8, 1890		-	1	23·2	1	17	18
,,	,,	15, ,,	-	2	27·3	2	36	24
,,	,,	22, ,,	-	3	23·0	4	48	29
,,	March	1, ,,	-	4	35·1	5	48	31
,,	,,	8, ,,	-	5	38·7	14	57	36
,,	,,	15, ,,	-	6	33·4	15	59	30
,,	,,	22, ,,	-	7	28·0	4	47	27
,,	,,	29, ,,	-	8	29·3	9	50	15
,,	April	5, ,,	-	9	23·3	8	30	27
,,	,,	12, ,,	-	10	23·7	4	30	19
,,	,,	19, ,,	-	11	31·6	5	44	22
,,	,,	26, ,,	-	12	29·1	3	46	19
,,	May	3, ,,	-	13	29·6	4	52	23
,,	,,	10, ,,	-	14	26·2	2	43	22
,,	,,	17, ,,	-	15	26·8	6	43	20
,,	,,	24, ,,	-	16	22·1	2	27	10
,,	,,	31, ,,	-	17	21·5	2	26	18
,,	June	7, ,,	-	18	23·7	1	26	16
,,	,,	14, ,,	-	19	21·0	1	36	8
,,	,,	21, ,,	-	20	24·4	1	28	10
,,	,,	28, ,,	-	21	19·6	0	22	13
	Mean	-	-	—	27·1	4·4	39	21

In the case of Sheffield it is evident there was no "sudden visitation."

Dr. Boobbyer,* Medical Officer of Health for the borough of Nottingham, gave a careful report of the epidemic which occurred in that town. Evidence of this sort is of the greatest value. Dr. Boobbyer made an analysis of the numbers affected in a total of 8,374 persons employed in various occupations, and gave a table showing the number of absentees from influenza during successive weeks in the first five months of 1890. In this table the fresh cases only are given. It is as follows:— *Nottingham, 2nd week in January.*

INFLUENZA. (FRESH CASES among 8,374 Persons employed in various Occupations.)

Date.	Number of new Cases.
Week ending January 18	26
,, ,, ,, 25	64
,, ,, February 1	160
,, ,, ,, 8	240
,, ,, ,, 15	308
,, ,, ,, 22	312
,, ,, March 1	274
,, ,, ,, 8	210
,, ,, ,, 15	154
,, ,, ,, 22	104
,, ,, ,, 29	78
,, ,, April 5	26
,, ,, ,, 12	22

* Lancet, 1890, Vol. II., p. 95.

From this table a chart (XII.) has been drawn out which shows graphically the rise and fall of the epidemic.

The height of the epidemic, it is seen, occurred during the week ending February 1st, and five weeks after the first statistics were collected.

In the case of Nottingham, as elsewhere, it is conclusively proved that there was no "sudden visitation," but that the number of cases which occurred, rose steadily day by day till the maximum was reached, and then as steadily declined.

Manchester, some cases in October. Dr. John Tatham, Medical Officer of Health for Manchester, has been kind enough to send me a copy of a most valuable report concerning the epidemic which afflicted that city. Dr. Tatham gives a most careful tabular summary of observations made on the atmospheric conditions which prevailed at Manchester during the period of the epidemic. This is not only interesting in itself but is an index of the care and labour which has been bestowed on the preparation of the report. With regard to the spread of influenza Dr. Tatham says that the first cases of influenza are said to have occurred as early as October, but the disease was not very prevalent till the end of the year 1889. In response to a circular which he sent to the medical men in private practice in Manchester, reports were sent in of the occurrence of 4,945 cases. The following table gives the dates at which the cases occurred, and the number of deaths.

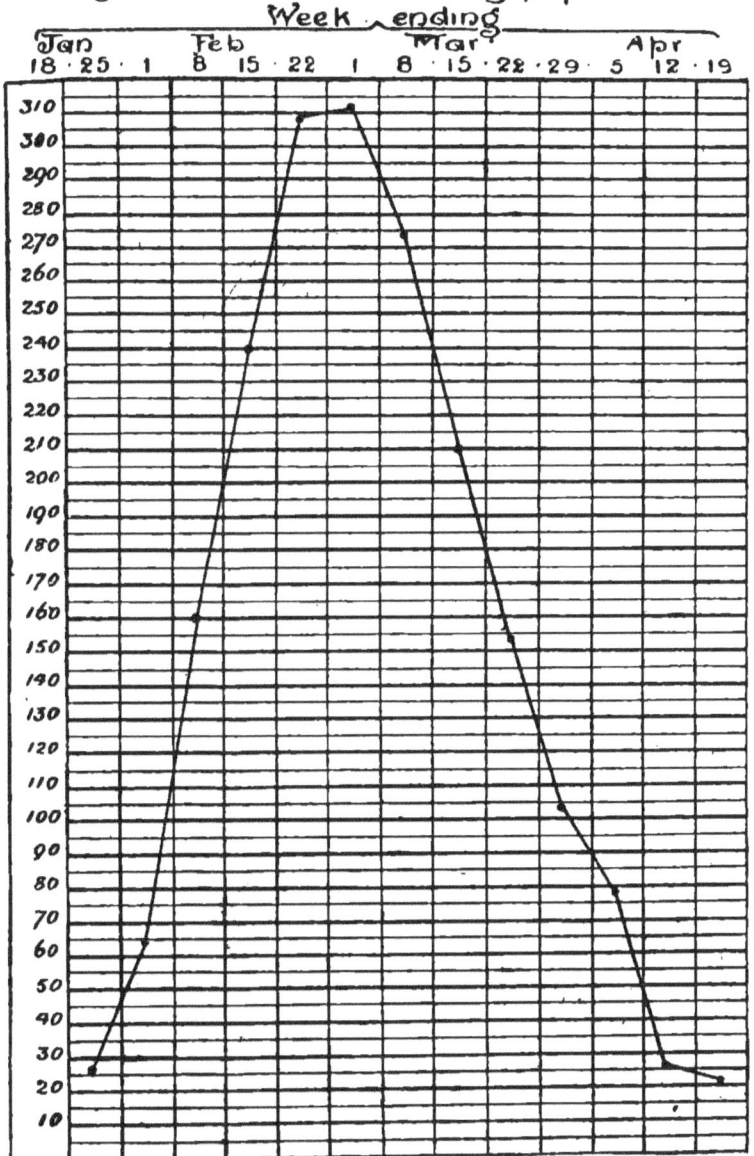

STATEMENT of NOTIFIED ATTACKS and of DEATHS from INFLUENZA in the CITY of MANCHESTER for each of the 25 Weeks from October 26th, 1889, to April 19th, 1890.

Date.	Number of Influenza Attacks.	Per-centage of Total Attacks.	Deaths from Influenza.
Week ending November 2, 1889	19	0·4	—
,, ,, ,, 9, ,,	31	0·6	—
,, ,, ,, 16, ,,	39	0·8	—
,, ,, ,, 23, ,,	43	0·9	—
,, ,, ,, 30, ,,	49	1·0	—
,, ,, December 7, ,,	94	1·9	—
,, ,, ,, 14, ,,	117	2·4	—
,, ,, ,, 21, ,,	81	1·6	—
,, ,, ,, 28, ,,	94	1·9	—
,, ,, January 4, 1890	177	3·6	—
,, ,, ,, 11, ,,	203	4·1	—
,, ,, ,, 18, ,,	233	4·7	1
,, ,, ,, 25, ,,	242	4·9	—
,, ,, February 1, ,,	316	6·4	3
,, ,, ,, 8, ,,	398	8·0	4
,, ,, ,, 15, ,,	487	9·9	7
,, ,, ,, 22, ,,	600	12·0	3
,, ,, March 1, ,,	539	10·9	5
,, ,, ,, 8, ,,	409	8·3	5
,, ,, ,, 15, ,,	290	5·9	6
,, ,, ,, 22, ,,	192	3·9	4
,, ,, ,, 29, ,,	107	2·2	2
,, ,, April 5, ,,	65	1·3	2
,, ,, ,, 12, ,,	69	1·4	2
,, ,, ,, 19, ,,	51	1·0	1

Even a casual glance at the table shows that influenza did not occur at Manchester as a "sudden visitation," but that the number of the cases increased week by week till they reached the maximum, February 22nd, and then diminished week by week.

The chart (XIII.) constructed from these figures shows this even more clearly. The height of the epidemic, it is seen, was during the third week in February. Chart XIV. shows the number of deaths.

Speaking generally, the inhabitants of rural districts suffered from influenza later than town dwellers.

Inverness, in February. Early in February influenza appeared in Inverness, and spread very rapidly.*

Stourbridge, epidemic 1st week in February. At Stourbridge cases occurred early in February, and between the 5th and 12th the inhabitants of Brierly Hill, a place near, became affected.

Berkshire and Dorset, epidemic middle of February. In the middle of February influenza was prevalent in Berkshire.†

At Wimborne (Dorset) influenza was very general about the middle of February, and the height of the epidemic at St. Ives (Cornwall) occurred at about the same time.

Rural parts of Derbyshire, epidemic in 3rd week in February. The rural parts of Derbyshire were much affected during the third week of February, and about the same time or a little earlier the disorder was spreading at Middlesborough (Yorkshire).

---

* British Medical Journal, 1890, Vol. I., p. 457.
† Northern Chronicle, February 12, 1890.

CHARTS XIII. AND XIV.

Chart shewing the weekly number of attacks of the Influenza in the City of Manchester during the epidemic there in 1889-90 for

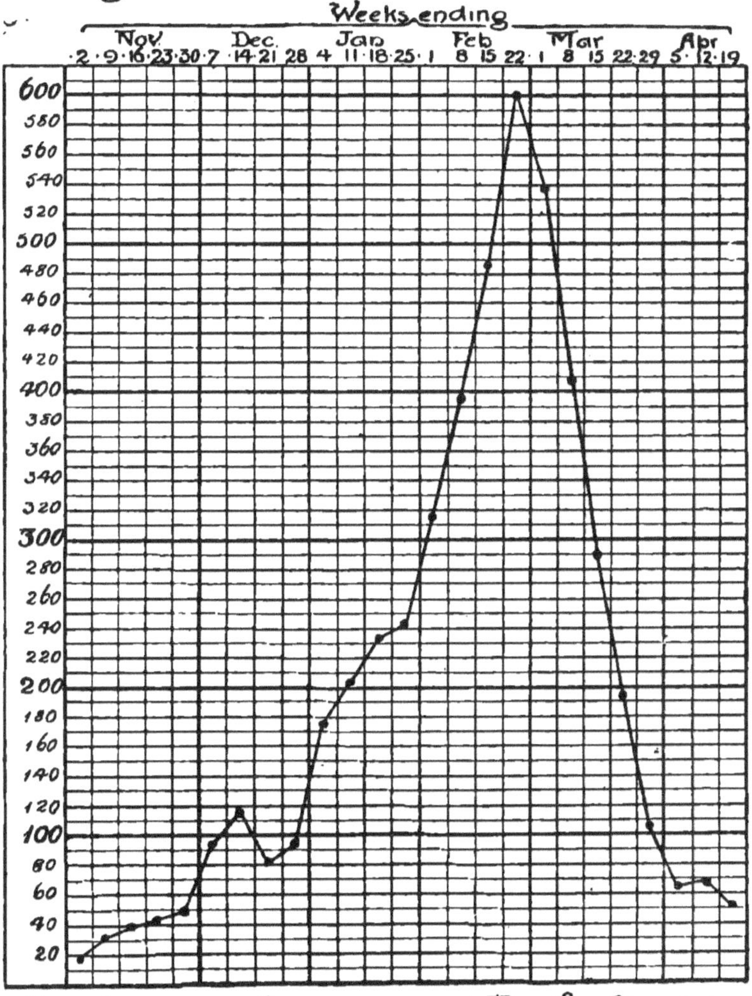

Chart shewing the number of Deaths from same for Weeks ending

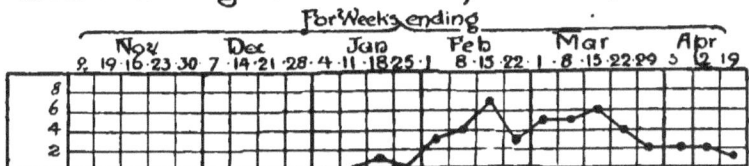

At Kilmersdon the pest was prevalent during the second week of February.

About the same time Dr. G. C. Taylor, Medical Officer of the Trowbridge district, found it necessary to close the public schools on account of the epidemic.

Trowbridge, epidemic 2nd week in February.

Darwen was affected towards the end of February, and the epidemic was at its height about March 1st.

Darwen, height of epidemic in March.

At the same date it was spreading in North Wales.

There were exceptions to the rule that towns suffered first. A remarkable instance of this occurred at Churchingford, in Devonshire, and I am indebted to Mr. W. J. Townsend Barker for some instructive details.

Churchingford.

Churchingford, he tells me, is on "an elevated tableland, situated almost equally distant from the towns of Taunton, Honiton, Chard, Ilminster, and Wellington, in Somerset, about seven to ten miles distant from each, and is elevated about 1,000 feet above the surrounding valleys on the north, west, and east, sloping gradually to the sea towards the south."

Mr. Townsend Barker is the only medical man in the whole district, which is very scantily populated. Influenza appeared in this retired spot before Christmas 1889. It is interesting to note how this happened. It was briefly as follows :—

On December 19th A. B. left Paris, where he had been staying at a hotel in which there were people

1st case in December imported from Paris, only one family and the doctor affected.

suffering from influenza. On the 22nd, after his return, he was taken ill, and had well marked symptoms of the disease. On January 2nd a daughter was seized, on the 5th a son was taken ill, on the 7th three other daughters. The next person affected was Mr. Townsend Barker himself. These were the first and, for a time, the only cases in Churchingford. There is no doubt that A. B. brought influenza with him from Paris, and his family and his physician were affected by contagion. It was a localised outbreak. "I believe," writes Mr. Townsend Barker, "that the reason why the disorder did not spread further was that I isolated both myself and the members of the family who were first attacked."

Churchingford, 2nd outbreak in February, imported from surrounding towns.
A second outbreak occurred later. On February 2nd a case of influenza occurred at Clayhidon, and the patient believed he had contracted it at Wellington, where it was then prevalent. On February 7th a case occurred at Bishopswood, in Otterford. In this instance it was apparently imported from Chard. During February the disease spread widely, and the epidemic reached its height about March 7th.

This series of observations must take rank with the historic ones of Dr. Haygarth, and of the distinguished French physicians, from whose writings I have already quoted. In England, as abroad, we again have impressed upon us the fact that in the case of a contagious disease, the isolation of the rural districts makes it easier to avoid the fallacies to which observations in large towns are inevitably exposed. In a country place all the people are, not infrequently, as in Mr. Townsend Barker's case, under the care of one man, who is, therefore, in a position to make certain of all the facts of such an outbreak as that recorded.

During the summer and autumn scattered cases of influenza were reported from Clifton,* Keynsham† (Somersetshire), Edinburgh,‡ and the City of London, but there was nothing like a general epidemic. Dr. Blake, of Great Yarmouth, tells me that he treated some cases during the winter of 1890-91, and I heard of a few others. A few scattered cases during winter of 1890-1891.

My friend, Dr. Daly, of Hull, tells me that he has no doubt that influenza was introduced into that town in 1891 from New York by one of the Wilson liners. The earliest cases Dr. Daly saw were sailors who had just landed. Most of the crew had been attacked on the voyage, and more than one had died. The disease had rapidly spread in Hull; but Dr. Daly remarks, "It was certainly localised to Hull for some days before spreading elsewhere." Outbreak in Hull, 1891, imported by American sailors.

Sheffield has also suffered severely in April 1891. An excellent report of the outbreak has been recorded by Dr. Theodore Thompson.§ Sheffield, outbreak 2nd week in April 1891.

"On this occasion," he writes, "the disease is supposed to have been imported from Hull, where it has for some time been prevalent. The outbreak

---

* British Medical Journal, 1890, Vol. II., p. 514.
† Ibid., 1890, Vol. II., p. 665.
‡ Ibid., 1890, Vol. II., p. 802.
§ Public Health, Vol. III., p. 420.

took place during the second week in April, when several persons were reported to be suffering from this affection, a report of which the accuracy was soon placed beyond doubt by the rapidity with which great numbers throughout the whole borough were seized with symptoms which corresponded to those usually attributed to influenza."

The epidemic which followed was more severe than the one which occurred in 1890, and the death rate rose greatly.

During the visitation a number of the inhabitants of Sheffield were called to London to give evidence before a Committee of the House of Commons concerning a proposed railway, which would have spoilt the beauty of one of the few quiet parts of the town. Several of the members of the Parliamentary Committee contracted influenza, but I am not aware that any careful record was made of the outbreak.

Cases of contagion. In concluding this chapter I now give some instances of the spread of influenza by contagion, and a brief account of two outbreaks which occurred at Haileybury College.

The following case has been recorded by Dr. Frank Clemow* :—

" A lady left a town in Shropshire, already invaded by the epidemic, on January 31st, and travelled to a village in Yorkshire, where there had not been a single case. On February 3rd she fell

---

* Public Health, Vol. II.

ill of the disease, and on the 5th, her little boy, who had travelled with her, was attacked. On the 8th, her mother and brother, who had never been in an infected district, but who had been in constant contact with the two other patients, were seized."

I am indebted to Dr. Charles Edward Shelly, of Hertford, for the two following cases:—

"The inmates of a large isolated country house were all in good health. In January 1890 a daughter of the house went to London to see her dentist, who was then so ill with influenza that he was scarcely able to stand, but he managed to perform various dental manipulations for more than half an hour. The girl came home feeling well; felt ill soon after getting home, and was quite collapsed at 10 p.m. The case was a severe one, and was followed by the illness of two sisters, who did not go away from home. These were the only three cases in that house for more than a month."

"A lady in good health, mistress of a large and healthy household, went up to London and spent the day in shopping. She was taken suddenly ill the following evening with severe influenza. This case was practically isolated for a fortnight, and no other cases developed. A large number of the staff at each of three of the establishments at which this lady called and spent sometime while in London, are known to have been then down with influenza."

I am indebted also to Dr. Shelly for some particulars of the outbreak of influenza which occurred *Outbreak at Haileybury College.*

at Haileybury College in 1890, and during the present year.

In 1890 the term began on January 24th, the first case of influenza occurred on January 28th, and on the 31st there were four fresh seizures. The greatest number of new cases in one day was 19 (February 14th).

From the material kindly supplied by Dr. Shelly a chart (XV.) has been made, which shows the number of new cases which occurred each week from January 28th to March 29th.

The second outbreak occurred in May 1891. The term began May 1st, but no case of influenza was seen till May 10th. Chart XVI. shows the numbers of the new cases which occurred each week from May 10th to June 20th.

It is evident that at Haileybury there was no "sudden visitation."

It would be tedious to review the whole of the evidence I have brought forward, but I wish to point out that wherever a careful record is given of an outbreak isolated cases of influenza preceded an epidemic.

It has been and still is taught that influenza occurs as a "sudden visitation," and it is argued from this that the disease does not arise from contagion, but from a general aërial contamination. In the epidemic of 1889–1890 I have been unable to find a single instance in which there was a sudden infection of a large number of people without the previous existence of isolated cases of the disease.

CHARTS XV. AND XVI.

Charts shewing the number of cases of Influenza which occurred at Haileybury College during the epidemics in 1890 and 1891.

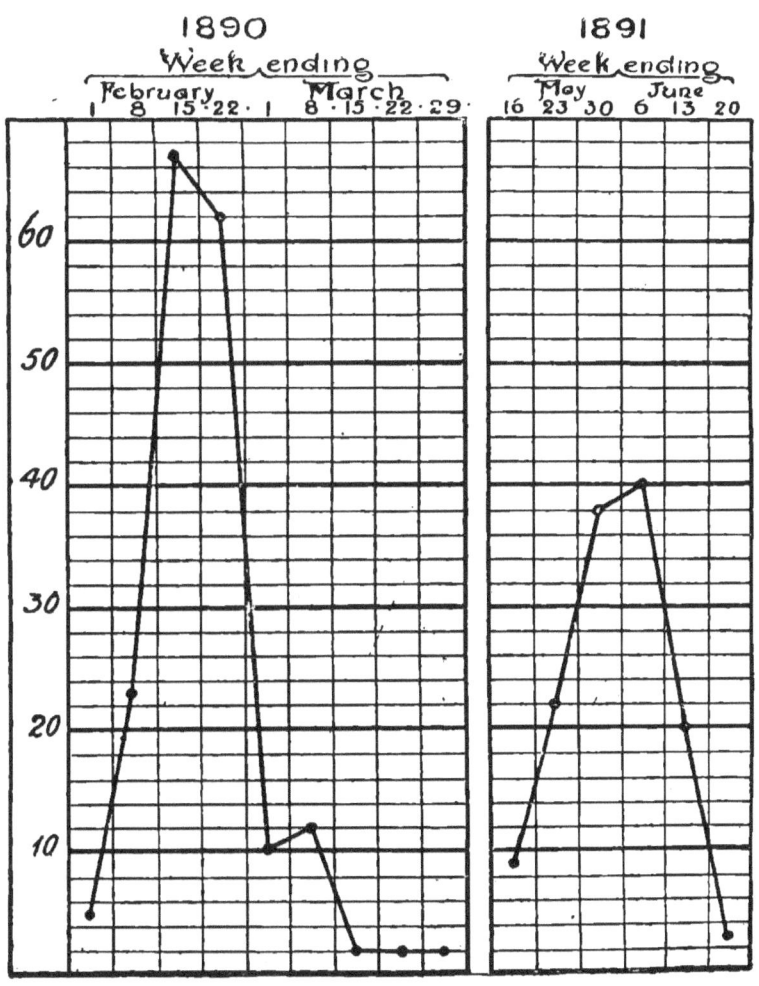

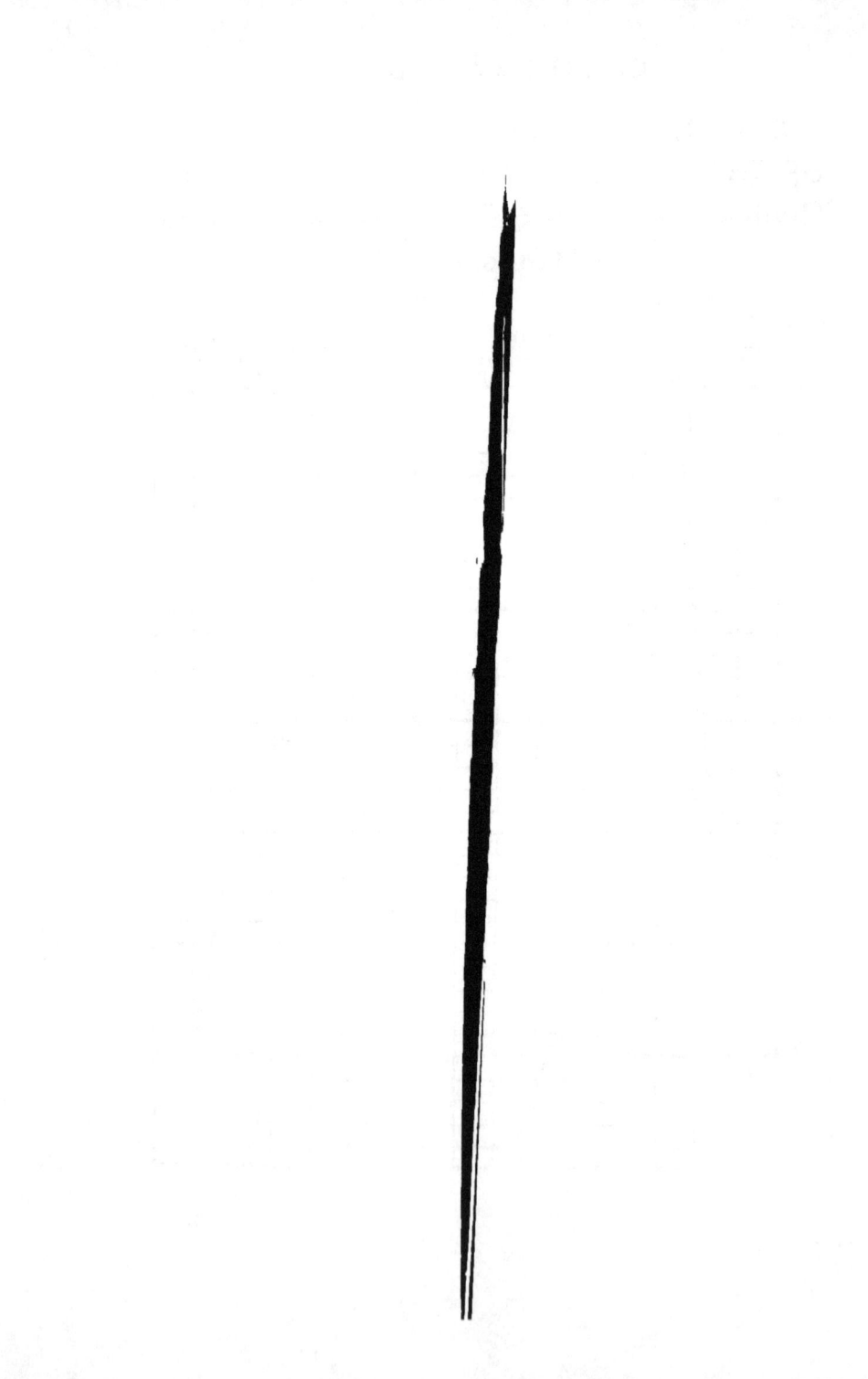

It is true that the first reports we had from St. Petersburgh in 1889, told of the sudden seizure of a large proportion of the population, but further investigation proved that influenza had for weeks existed in the Russian capital.

In England no sudden visitation was reported from any part of the country. The same story came from towns and the rural districts. "There is no epidemic of influenza here, a few scattered cases have occurred." This was the first report, later it was said, "The epidemic is here, and is spreading into the villages round." In many instances the first patient seized had come from an infected place.

A glance at the charts in this chapter shows that the same course was taken by the disease in every instance. The number of the casualty patients at St. George's rose gradually and gradually fell. The London death rate rose with the spread of influenza, and dropped as the epidemic subsided. The number of absentees from illness at Broadwood's workshop rose gradually and gradually fell. At the Oxford University Press exactly the same thing happened. Careful investigations at Nottingham proved that there was a gradual rise in the numbers affected till the maximum was reached, then a gradual fall. At Manchester the records showed the same course of events. At Haileybury College in 1890 and 1891 influenza ran the same course as it did elsewhere.

In every case where the course of the disease was studied with care it was seen that it spread in the same way as any other contagious disease.

Another point worthy of notice is that influenza spreads along the most frequented lines of human intercommunication. It is obviously impracticable to trace out all the possible routes of infection taken by such a disease in its entrance into and travels in a country like England, with its multitudinous inlets and lines of communication. When these facts are borne in mind it will be readily acknowledged that the unexplained appearance of the disease in an isolated place only points to our incomplete knowledge as to its introduction there, and need never excite wonder.

The broad outlines of the course of its travels are, however, known to us. London, according to Dr. Buchanan, was affected as early as October 1889, and Manchester as early as November. Colchester, Canterbury, Chelmsford, Oxford, and Liverpool were affected early in January 1890, and Birmingham in the second week of that month.

Not until February did influenza appear at places so remote as Inverness, Stourbridge, and Wimborne. In the third week of February the disease was prevalent in the rural parts of Derbyshire, and as late as March influenza was spreading in the sparsely populated parts of Wales.

A remarkable isolated outbreak occurred at Churchingford (Devon) as early as December 1889, when a gentleman who returned home from Paris was seized with the disease, and gave it to his family and to his doctor. In February and March 1890 Churchingford was again visited, but this time the infection of the disease was distinctly traced in more than one instance to a town where it was then prevalent.

According to the "aërial contamination" theory, it is impossible to conceive how it is that influenza does not affect small villages in its course through the air from one town to another, or why villages should be affected later than towns.

"This slowness and irregularity in the progress of the disease may be easily accounted for, if we admit that it was conveyed by infected persons; but if we suppose it was conveyed by the air, they seem utterly inexplicable."*

In fact, to put the matter briefly, the facts observed with regard to the course of influenza show that it is spread by contagion.

---

* Dr. Edward Gray, Medical Communications, Vol. I.

CHAPTER X.

PRISONERS AND THE INFLUENZA EPIDEMIC.

During the epidemic of 1803 the inmates of Ryegate (Reigate) Workhouse, Hereford Asylum, and Worcester Jail escaped the epidemic. In 1889 the nuns of Charlottenbourg did not have it, although it was prevalent in the town. At the Royal Morningside Asylum isolated cases preceded the epidemic. The prisoners escaped at Bodmin, Ipswich, Kendal, Knutsford, Lewes, Norwich, and Portsmouth, although influenza was raging in the towns in which the prisoners were situated. These facts cannot possibly be explained by the doctrine of " aërial contamination." Prisoners are not deprived of air, but they are kept more apart from human contact than other people, hence they are less liable to the chances of contagion. The records of the Scottish prisons show that in none of them was there a " sudden visitation." In each prison in which an epidemic occurred a considerable time elapsed between the first appearance of the disease and the height of the epidemic.

The epidemic of 1803. We have seen how numerous are the instances in which the introduction of influenza into an uninfected place has been traced to an individual who has carried the disease with him,

It has also been noticed that isolated cases of influenza have preceded a general infection, and that towns are affected before the villages around.

We have now to see how influenza affects those who are kept apart from human intercourse by poverty, crime, or religion.

During the epidemic of 1803 the inmates of Ryegate* Workhouse escaped. This was at the time thought to be due to the fact that many of the paupers were employed in making blankets, and used oil in the manufacture.

In the same year the lunatics in the Asylum at Hereford† did not have influenza, and at Worcester‡ the Governor of the prison and a daughter had it, but the prisoners were spared.

Many similar instances could be quoted from old records.

During the epidemic of 1889–1890, the prisoners and nuns escaped influenza when it was raging in the town in which their prison or convent was situated. The following case was reported from Germany.§ At Charlottenbourg there is a convent in which there are over a hundred women, who are weak and ill, and some of whom are afflicted with pulmonary disease. The inmates of this convent never cross its portal. Their rule is strict; communication with the external world is made by women, but only in an indirect way. Only two men, the priest and the doctor, have direct relation

marginal note: The nuns at Charlottenbourg escaped influenza although it was prevalent in the town.

---

* Medical and Physical Journal, Vol. IX., p. 580.
† Ibid., Vol. X., p. 127.
‡ Mem. Med. Soc., Vol. VI. pp. 435, 436.
§ Semaine Médicale, 1889, p. 471.

with the inhabitants of the convent; and although influenza was prevalent at Charlottenbourg, no one in the convent had it.

<small>At the Morningside Asylum, Edinburgh, isolated cases preceded an epidemic.</small>

The inmates of the Royal Asylum, Morningside, Edinburgh, suffered from a great outbreak of influenza. A report of this was published by Messrs. George M. Robertson and Frank A. Elkins.*

The report says, "During the second and third weeks in December, scattered cases of illness, which we did not at the time differentiate from ordinary colds, occurred, but during the fourth week of the month we became certain that the epidemic was amongst us by the daily invasion, by the number attacked, and by the nature of the symptoms." From this statement I think we may conclude that isolated cases occurred before there was a general epidemic. It is very much to be regretted that no record was kept of the numbers of new cases affected daily. Such facts would have been of the greatest interest, and might easily have been recorded.

<small>Influenza in English prisons.</small>

The Thirteenth Report of the Commissioners of Prisons gives some account of the way in which the epidemic afflicted prisoners.

<small>Birmingham.</small>

From Birmingham "the medical officer reported that no diseases requiring notification under the Contagious (Infectious) Diseases Act" happened during the year which ended March 31st, 1890. In the next paragraph it is stated:

"The epidemic of influenza occurred in the prison during January."

---

* British Medical Journal, 1890, Vol. I., p. 228.

The report from Bodmin jail was as follows:— *Bodmin, prisoners not affected.*
"There has been no epidemic among the prisoners, but most of the officers have suffered from influenza."*

At Canterbury, where, as we have seen, influenza was very prevalent, the disease was introduced into the prison by a patient.† *Canterbury, influenza introduced by a patient.* "The general health of both divisions of the prison has been good. The medical officer reported that there had been nine cases of influenza, and that ten officers were affected with the same epidemic about the same time. The death of a prisoner had occurred from acute inflammation of both lungs, accompanying an attack of influenza caught before admission into the prison."

It is interesting to note that the number of officers affected was greater than that of the prisoners.

The Exeter prison report was as follows:— *At Exeter it affected the prisoners later than non-resident officers.*
"The malarious complaint denominated influenza, rife in the neighbourhood of the prison, as everywhere else, invaded the houses of the non-resident officers, and very late made its appearance within the walls of the prison. The cases were, however, not numerous, speedily yielded to treatment, and were uncomplicated."‡

On the theory that influenza is spread by aërial contamination, it would not be easy to explain its

---

* Thirteenth Report of the Commissioners of Prisons (for the year ended 31st March 1890), London, 1890., p. 9.
† Op. cit. p. 15.   ‡ Op. cit. p. 31.

late appearance in Exeter prison. If influenza be spread by contagion the matter presents no difficulty. It is evident that the non-resident officers had an opportunity of taking the disease by contagion in their intercourse with the outside world. Such a chance of infection was denied to the prisoners.

At Ipswich, Knutsford, and Kendal, the prisoners escaped influenza.

At Ipswich jail there was no epidemic,* but influenza affected great numbers of the people of the town.

The report from Kendal tells exactly the same tale :—

"The medical officer reported that sanitary condition of the prison and the health of the prisoners had been exceedingly good. There had been an entire freedom from all disease of an epidemic character, and no case of influenza occurred in the prison, notwithstanding its general prevalence in the town of Kendal."†

At Knutsford, too, the prisoners escaped. The report says :—

"The sanitary condition of the prison has been satisfactory, and the general health of the prisoners has been good. There has been no case of infectious disease."‡

At Liverpool, Leicester, and Maidstone there were outbreaks.

At Leicester§ there were a few cases of influenza in the jail.

---

* Op. cit. p. 39. † Op. cit. p. 40.
‡ Op. cit. p. 44. § Op. cit. p. 50.

No "aërial contamination" affected the prisoners at Lewes. The prison report says:—

"There has been no epidemic disease, although influenza was very prevalent amongst the officers and their families at the beginning of the year 1890."*

At Liverpool† and at Maidstone‡ there were outbreaks.

At Norwich the prisoners escaped. *(At Norwich, Lewes, and Portsmouth the prisoners escaped influenza.)*
"The officers (with two exceptions) have all suffered from the prevailing epidemic of influenza, but there was no epidemic in the prison, and the water and drainage are good and satisfactory."§

At Portsmouth exactly the same thing happened. The report from the jail was:—
"There has not been a single case of influenza, although this complaint has been very prevalent in the town."‖

At Preston¶ "many officers and prisoners were afflicted with it." The numbers are not stated. *(At Preston and Reading there were epidemics.)*
At Reading** "about 25" had influenza.

From Ruthin†† prison:—
"The medical officer reported that, although influenza had been very prevalent in the locality, only one prisoner took the disease." *(At Ruthin jail there was one case.)*

At St. Albans‡‡ "several prisoners" were affected.

---

* Op. cit. p. 52.   † Op. cit. p. 55.   ‡ Op. cit. p. 58.
§ Op. cit. p. 67.   ‖ Op. cit. p. 76.   ¶ Op. cit. p. 78.
** Op. cit. p. 80.   †† Op. cit. p. 82.   ‡‡ Op. cit. p. 84.

PRISONERS AND THE INFLUENZA EPIDEMIC.

At Strangeways the prisoners were affected.

The report from Strangeways* was as follows:—

"The medical officer reported that during the early part of 1890 a somewhat severe outbreak of Russian influenza occurred in the prison, especially on the female side, nearly 100 prisoners and officers were attacked, but fortunately no case proved fatal."

A more complete account of this outbreak was given by Dr. Tatham, the Medical Officer of Health for Manchester, in a "Memorandum on Influenza Prevalence in Manchester." I have already had occasion to refer to this valuable record, which is in every way a model of what such a report should be:—

"Her Majesty's Prison in Strangeways suffered somewhat severely," it is stated, "from this disease during the month of February, a few subsequent cases occurring about the middle of March. From the 1st to the 15th of February influenza threatened to become very troublesome, for the prison contained at that time an average population of 1,000, and of these not less than 57 persons were attacked during that fortnight alone. The total number of persons affected at the prison during the epidemic was 77."

At Wandsworth jail there was an epidemic.

At Wandsworth there was an epidemic in the jail:—

"In the month of January the influenza epidemic, which was so prevalent throughout London, attacked officers and prisoners to the number of 180."†

---

* Op. cit. p. 91.   † Op. cit. p. 99.

From Winchester the report was as follows :— *At Winchester prison there were three cases.*

"There was a severe outbreak of epidemic influenza among the warders' families, but the warders, without exception, escaped the disease, and in the prison three prisoners only were attacked."

The many instances in which prisoners escaped influenza is too remarkable to escape attention. The report from which I have quoted has been fairly treated. In each case in which influenza has been mentioned I have mentioned it. I have not picked out cases to prove any theory, but have quoted all the facts mentioned which bear on the subject.

The following is a summary of the chief points of interest :—

I.—Prisons in which Outbreaks of Influenza occurred.

Birmingham, (numbers not given).
Canterbury, (imported by a prisoner), 9 prisoners and 10 officers affected.
Exeter, cases not numerous, numbers not given, non-resident officers affected first, the disease appearing in the prison very late.
Preston, many officers and prisoners affected.
Leicester, a few cases.
Liverpool.
Maidstone.
Reading, about 25 prisoners affected.
Ruthin, 1 prisoner affected.
St. Albans, several prisoners.

Strangeways, a severe attack, 100 officers and prisoners.
Wandsworth, 180 officers and prisoners.
Winchester, 3 prisoners.

II.—Prisons in which there was no Influenza, although it was raging in the Town.

Bodmin.
Ipswich.
Kendal.
Knutsford.
Lewes, very prevalent amongst the officers and their families.
Norwich, all the officers but two affected.
Portsmouth.

Thus the prisoners in seven out of a total of twenty jails entirely escaped infection, although influenza was prevalent in the town.

In those records in which any information is given on the subject, the evidence shows the strong probability of the theory that the officers spread the disease, or that it was imported by a prisoner. Thus at Exeter the prisoners were affected after the officers, and at Canterbury a prisoner took influenza in with him.

It is much to be regretted that the numbers of prisoners affected, and the dates at which they were affected, have not been given in the report.

The chief question which arises from a consideration of the Report is: How was it that the inmates of seven prisons escaped influenza, when it was raging in the town in which these prisons were?

Before considering the significance of these striking facts it will be convenient to see what occurred in Scotland during the presence of the same epidemic.

The Twelfth Annual Report of the Prison Commissioners for Scotland contains some information about the numbers of prisoners and of officers affected, the date at which the first case was noticed in each place, the date of the height of the epidemic in each place (except where there were only two cases at Stornoway), and the date of the disappearance of the ailment.

Sir Douglas Maclagan, M.D., gives a table in which these points are shown at a glance. (*See* page 100.)

A study of the table proves that it took some time for the epidemic to reach its height both at Perth and at Glasgow.

In prisons, as elsewhere, isolated cases precede an epidemic.

There was not in the case of any prison a sudden sickness of a large number of prisoners, and no more "occult influence" than contagion is necessary to explain the spread. It is much to be regretted that no daily, or even weekly, statistics, are given of the numbers of the prisoners and officers who were attacked. Such statistics, in the case of prisons, are made with little trouble, and are of immense importance in the study of such a disease as influenza.

ABSTRACT of RETURNS as to the EPIDEMIC of INFLUENZA which occurred during the year ended 31st March 1890.

Prisons.	Date of first Appearance of Influenza.	Date when Ailment reached its height.	Date of Disappearance of Ailment.	Number of Officers ill.	Number of Prisoners ill.	Average Number of Days' Illness.	
						Officers.	Prisoners.
Dundee	9th January	11th and 12th January	8th February	2	7	8½	6½
Glasgow	19th December	20th January	5th „	7	11	5	5
Greenock	23rd „	17th February	3rd March	5	4	8	5
Inverness	13th February	16th „	3rd „	—	3	—	1
Lerwick	11th „	14th „	22nd February	3	—	10	—
Lochmaddy	9th January	7th „	21st „	1	—	21	—
Maxwelltown	18th March	19th March	23rd March	1	—	5	—
Perth (General)	7th January	22nd February	24th „	13	59	13	7
Peterhead (General Convict)	12th „	8th March	7th April	7	64	8	10
Stornoway	14th February	3rd day after declaring itself in each case.*	20th March	2	—	10	—
				41	148		

* The medical officer to the prison at Stornoway apparently misunderstood the nature of the question he was asked.

It is only fair to add to this suggestion and criticism, that the report from Scotland is more complete and more carefully made than the corresponding English report, which has just been considered.

In his comments on the epidemic, Sir Douglas Maclagan says :—

"The epidemic of influenza cannot be said to have invaded the prisons extensively, except at Perth and Peterhead. It is worth while noticing how many of the officers were attacked in comparison with the prisoners—41 of the former, as against 148 of the latter."

Now, what is to be learnt from this? How was it that prisoners escaped their fair share of influenza?

It has been argued that the retired life of the prisoner has enabled him to avoid it, because he was not exposed to the air. It is curious that this explanation has been lately revived. The fallacy of it was exposed nearly a hundred years ago by Dr. Gray,* who said :—

"The argument that those most exposed to the weather were generally the first persons attacked, is surely by no means in favour of the opinion that the cause of the disorder resided in the air; for if it had resided there, what should have prevented those who staid at home from being infected; since the air they breathed must necessarily have been the same as that breathed by those who went out; but if, on the other hand, a com-

---

* Medical Communications, Vol. I.

munication with some infected person was necessary to produce the distemper, it is very clear that those who went out of doors and mixed with the world were more likely to get it, than those who did not stir from home."

The reason, therefore, why the prisoners escaped, was because they did not stir out, and they thus escaped contagion. A general aërial contamination could not be so easily avoided, even by a prisoner.

CHAPTER XI.

THE SPREAD OF INFLUENZA BY PARCELS.

In 1782, Dr. Mease recorded his conviction that he received the infection of influenza from a parcel of wearing apparel. In 1889, a case was recorded, which pointed to the possibility that the disease might be conveyed in packages. In 1890, at Alderney, a Custom House officer was the first person affected; it was conjectured that he might have derived the infection from parcels, and another case was recorded, in which a housekeeper was seized with influenza a week after she had received a present, which had been on the bed of a patient who had the disease. In 1880, an outbreak occurred after the exposure of the embalmed body of a man who had died from influenza, but it was not clear that there was any connexion between the body and the outbreak of the disease. In 1889, a patient took influenza, after being put into a bed which had before been occupied by a patient who had the disorder. The spread of influenza by parcels is not proved by any of these cases.

A few observations have been made which support the view that infective material of influenza

Epidemic of 1782.
Case of

<small>*infection possibly derived from wearing apparel.*</small> can be carried by parcels. Dr. Mease wrote in 1782:—*

"I have no shadow of doubt that the disorder was contagious, and am certain I myself received the infection from a small trunk of wearing apparel which came from Dublin, where it then raged. I may add that this was the first introduction of it into the town."

Dr. Gray† attached some importance to this statement, and rightly thought that contagion by parcels was more probable than a general aërial contamination. He wrote as follows:—

"Some, who had no doubt that the disorder was communicated by infected persons, yet thought that it might also be conveyed to a *considerable distance* by the air. This latter opinion, however, certainly cannot be supported by any analogical reasoning from known facts; for though the absolute contact of an infected person is not supposed necessary to convey a contagious disease, we have no reason to think the power of communicating it extends to any considerable distance in the open air; a free exposure to which seems so to divide, and dilute (if the expression may be allowed) all infectious effluvia, that their virulence is entirely destroyed. Even the infection of the plague is not supposed to be communicable to any great distance by the air alone, though it is admitted, that certain substances such as cotton, wearing apparel, &c., may be impregnated with it, and if secluded from the air, may convey the disorder from one place to another; in

---

* Medical Communications, Vol. I.
† Medical Communications, Vol. I.

one instance the influenza seems to have been transported in that manner."

Dr. Danguy des Déserts has recorded the following case :—*

"On December 11th an officer received from Paris two large packages which came from the house of Potin. They were contained in boxes covered with wood shavings. He did the unpacking himself. Three days after he was seized with influenza. The next day and the day after his wife and three servants took it. I believe I am right in saying that these cases are the first which had been seen in Brest." *[marginal: Epidemic of 1889-1890. Case of infection possibly derived from packages.]*

This case is not altogether satisfactory, but is still worth noting. Unfortunately Dr. Danguy des Déserts does not seem to have been certain that the officer was the first person affected in Brest, nor is this uncertainty to be wondered at, when one remembers the size of the town.

The next piece of evidence I have to offer is from Alderney, the isolated position of which makes such observations easier to make and also more reliable.

Writing on January 25th, 1890,† Mr. Bernard said :—

"It might be interesting to some of your readers, inasmuch as this island is only in communication with Guernsey and Cherbourg, to know that the epidemic is amongst us. I have had six genuine cases since January 14th, and I think that the garrison localised here is also suffering from it. As *[marginal: Case of infection possibly derived from parcels.]*

---

* Semaine Médicale, 1890, p. 5.
† Lancet, Vol. I., p. 376.

far as I could I have made sure that none of the passengers landed ever had the epidemic, either during their stay here or before their coming. The first case was a Custom House officer, whose duty is to take note of all the parcels coming from England *via* Guernsey. I may also state that the mail boat is the only one which has been here for weeks."

There is a certain element of doubt even in this record, and a scientific fact cannot be accepted on any but the most positive evidence.

Dr. Bezley Thorne* has recorded, evidently on hearsay evidence, a somewhat similar case. In a letter to the "Lancet" he said :—

Case of infection possibly derived from Christmas presents.

"As the question of the infectiousness of influenza is at the present time naturally exciting attention, I think that the following statement may be read with interest :—On December 22nd last a patient of mine, who, ten days previously, had commenced making daily visits to the sick room of a relative who had fallen a victim to the epidemic, herself developed the initial symptoms of a sharp attack. On the following day a number of presents, which had been selected for the domestics of her country establishment, were placed on her bed, and she herself folded some of them in paper. On the eighth day after their delivery the housekeeper and two of her children fell ill of influenza, as well as two other servants who had received some of the gifts. A week later the malady, as I am informed, made its first appearance in the village."

---

* Lancet, 1890, Vol. I., p. 138.

It seems a pity that this outbreak was not more carefully inquired into, the evidence as it stands is of little value.

Dr. John Guiteras and Dr. J. William White have made a very interesting contribution to the history of the spread of influenza.* The record contains notes on nineteen cases. The first occurred in Europe in December 1879, and the patient died, his body was embalmed and taken to America, where it was exposed to the view of his friends. A daughter, son, and nephew of the deceased sailed from Liverpool on January 3rd, 1880, and on January 19th the son (case 2) began to be ill, his chief symptoms being headache, earache, cough, and pharyngitis. On February 8th a lady who visited the last patient (case 3) was seized with a feeling of weakness which was followed by headache, cough, and much sweating. (Case 4.) On February 10th a sister of case 2 had headache and vertigo, this was followed by fever and sweating, and there was much cough.

Cases of infection possibly derived from an embalmed body.

Ultimately the number of cases reached twenty or twenty-one, but the disease did not spread " beyond the immediate circle of those who were in direct communication with the invalids."†

The writers of the paper ask, " Did the patients affected in our city contract the disease from the first case, or did they get it from the remains which

---

* Philadelphia Times, April 10, 1880.

† A System of Practical Medicine, by American Authorities. Edited by William Pepper, M.D., LL.D. Assisted by Louis Starr, M.D., Vol. I., p. 863. London. Sampson, Low, Marston, Searle, and Rivington, 1885.

were embalmed, brought here, and after exposure interred?"

The paper is an interesting one from several points of view,* and the whole outbreak was a remarkable one. There was, however, a possibility that the spread of the disease was due to contagion, and it is by no means certain that the embalmed body did more than lend an element of romance to the whole story. I must refer the reader to the original paper for full particulars.

Professor Bäumler,† in the pamphlet to which I have already referred, mentions, but does not quote, some observations of Professor Schauta, which he thought proved that the contagion of influenza can be conveyed by fomites. The report is to be found in the "Prager Medicinische Wochenschrift."‡ It does not appear to me to be very convincing.

Case of infection possibly derived from a bed.

Professor Schauta pointed out that influenza occurred in the town before it affected anyone in the lying-in hospital. The first person seized was a head-nurse, the second an attendant. Professor Schauta, to prevent the spread of the disease, had taken the precaution to have a separate room prepared for such an emergency, and the attendant was removed thither. No new case of influenza occurred, and this room was then occupied by unaffected patients.

---

* Note especially the remarks on the physical signs in the lungs of those affected. Loc. cit., p. 339, 342.
† Ueber die Influenza von 1889 und 1890.
‡ 1890, Vol. X., p. 125.

Now, however it turned out that in this room the patient who occupied the bed, formerly occupied by the attendant, was the first to take influenza, then the two women nearest her, while the remaining patients were wholly spared. (It is not clear here whether Professor Schauta is referring to others in same room or in other parts of the hospital.)

Briefly, the evidence in favour of the spread of influenza by parcels is not convincing, and facts bearing on the subject must be recorded with the greatest care, if they are to be of any value.

CHAPTER XII.

THE INCUBATION PERIOD OF INFLUENZA.

Difficulties in determining the length of the incubation period of influenza. It may be less than the time which elapsed between the first chance of infection and the first symptoms. According to M. Antony, the incubation period in his cases varied from one to four days. In Dr. Bordone's cases from two to five days. In the cases recorded by Dr. Tueffart it was four days in two cases and five days in one case. In two cases reported by Dr. Védel it was only a few hours. Dr. Delépine has recorded two cases in which the incubation period was less than twenty-four hours.

On board a ship at sea, or on a Light-ship, there is no possibility of infection until the crew are exposed to direct contagion, either when the ship lands at a port or takes someone on board. In such cases the maximum time of the incubation period can be determined. Two such cases have come to my knowledge. Cases of a sailor on the " Correo " and of the mate on the Outer Dowsing Light-ship.

The incubation period of influenza. An important factor in causing the rapidity in the spread of influenza is the shortness of the incubation period of the disease. Dr. Antony,* from his experience at the Val de Grâce Hospital, came to the conclusion that the incubation varied considerably. He stated it to be :—

may vary from 1 to 4 days, according to M. Antony.

1 day in 2 cases,
2 days in 4 cases,
3 ,, 4 cases.
4 ,, 1 case.

There must necessarily be some doubt about the absolute correctness of any such observations on this subject, for it does not follow that a patient is always affected on his first exposure to infection. On the other hand, it may sometimes be said with certainty that the incubation period has not exceeded a certain time. This happens when we know the moment at which a patient was first exposed to infection, and the time at which the first symptoms presented themselves.

It is often impossible to determine the incubation period.

In the case of the outbreak at Frontignan, which was reported by Dr. Bordone,† it may be remembered that a sufferer from influenza gave a dinner party on December 17th, and five of his guests had symptoms of it on the 19th; that is to say, the incubation period was not more than two days. The host on the 18th went to his office, and on the 21st his *employé* had influenza, that is three days after the first possibility of contagion. The *employé's* mother did not have any symptoms till five days after her first exposure to infection.

Dr. Bordone's cases apparently had an incubation of from two to five days.

---

* Bulletins et Mémoires de la Société médicale des hôpitaux de Paris. Ser. 3, Vol. VII., 1890, p. 95.
† *See* p. 56.

In the series of cases reported by Dr. Tueffart* the daughters of the first sufferer had their first symptoms four days, and the son six days after their father was seized.

<small>Dr. Védel reported cases which had an incubation period of a few hours only.</small>

It would be tedious to go over all these histories, and I will only refer to the one reported by Dr. Védel.† The gentleman of Vergèze who visited the afflicted family at Lunel and remained with them for two hours, was seized with influenza the same night, and his wife had it the next day. It is clear from these cases that the incubation may be one of a few hours only.

<small>Dr. Delépine published instances in which the incubation period was under 24 hours.</small>

Dr. Delépine, in the paper to which I have already referred,‡ has given some very careful notes on two cases which may be here quoted.

"*Case* I.—A lady, Mrs. Y., who had during the previous six weeks, been afflicted with feverish colds, with marked catarrhal symptoms (to which she is very liable), was quite well again, when she received the visit of Mr. A., who remained for about two hours with her on that day. Mr. A. (ten days previously) had had a very severe attack of influenza, with high fever and great prostration. this gentleman was the first person having had influenza that Mrs. Y. had seen.

"On the 19th Mr. A. called again on Mrs. Y. In the evening of the same day (5 P.M.) Mrs. Y. began to feel very heavy and depressed, and to move about with difficulty. At 8 P.M. she felt intensely cold, although her face was flushed, and her skin

---

* *See* p. 56. † *See* p. 58.
‡ "Is influenza a contagious or a miasmatic disease?" Practitioner, Vol. XLIV.

felt burning to the touch. At 10 p.m. the temperature was 104°," and to be brief she had influenza.

" *Case* II.—Mrs. Y.'s daughter Z., a little girl of five, was quite well on January 21, and had not seen her mother since the beginning of the latter's illness. On that day another little girl, a year older, came to stay with her in the evening. That little girl had had (six days before) a sudden febrile attack, a feverish cold, which according to two medical men who saw the case, was attributable to influenza.

"*January* 21.—At 8 in the evening Mrs. Y.'s daughter first saw her little companion. She was then quite well and happy . . . . .

"*January* 22.—In the morning the children began to play together. In the evening Z. showed signs of not being well. During the night she slept badly, waking up suddenly as if greatly frightened, and under the influence of great cerebral excitement (a thing quite unusual to the child under any circumstances; she had often had the same little friend with her before).

"*January* 23.—8 a.m., her face was pale and pinched; her eyes red; the tongue furred; the cervical glands slightly enlarged. There was no cough. The child complained of feeling cold and sleepy, that her eyes felt as ' if she was going to cry.' She also complained of frontal headache, and of pains in the back;" and in short had symptoms of influenza. In these cases the incubation period was less than twenty-four hours.

In the case of a sailor on board the Norwegian barquentine "Correo" the incubation period cannot have been more than three days.

Through the kindness of Professor Wynter Blyth and of Dr. Collingridge, Medical Officer of Health for the Port of London, my attention was called, in May 1891, to an outbreak of influenza which occurred on board the Norwegian barquentine "Correo." This vessel entered the West India Dock on May 10th from Minititlan, and the crew eight in number had enjoyed good health throughout a voyage of 10 months to various unhealthy ports in South America. Soon after her arrival in London, everyone on board had influenza. They were affected in the order given below. The plan of the Correo (Pl. XVII.), which was kindly made for me by Mr. Henry Spadaccini, shows the arrangement of berths, and the numerals on the plan show the order in which their occupants were affected.

1. Sailor seized May 12th.
2. Captain „ 14th.
3. 1st Mate „ 17th.
4. 2nd Mate „ 17th.
5. Boy seized „ 18th.
6. Steward „ 19th.
7. Boy „ 19th.

The first possibility of infection in this case occurred on May 9th, when a pilot went on board in the Downs. On the 10th the vessel was in dock and many people boarded her. In the case of a sailor who was taken ill on the 12th, the incubation period cannot possibly have been more than three days, and was probably not more than two. The fact that every one on board took

PLATE XVII.

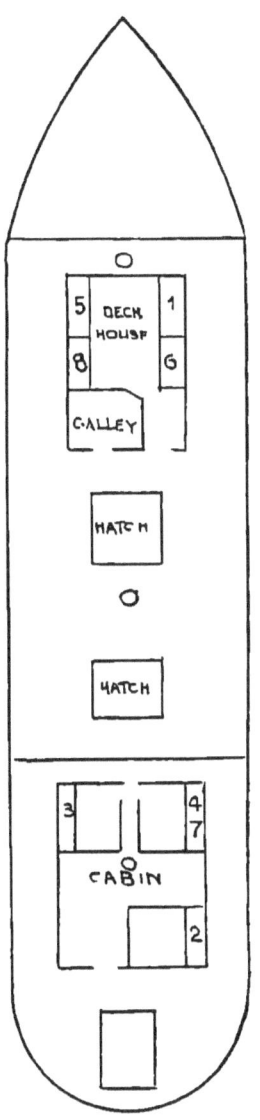

BARQUENTINE "CORREO" OF CHRISTIANSAND.

influenza confirms observations which have been previously made, that healthy people arriving from a distance at an infected place are particularly liable to take the disease.

I am indebted to my friend Dr. Blake for sending me some particulars of a very interesting case, and for assisting me to investigate the matter with him. The mate on the Outer Dowsing Light-ship was relieved from duty on April 15th, but before he arrived in Yarmouth on the 16th, he developed symptoms of influenza. There had been no illness on the Light-ship, and there was no history of any on the Trinity House boat, which relieved him. From the account first received it might have been supposed that the man contracted influenza whilst on the Light-ship.

In the case of the mate on the Outer Dowsing Light-ship the incubation period cannot have been more than 24 hours.

Dr. Parsons* has already mentioned the fact that in the epidemic of 1889–1890 the men on those ships had not been known to have influenza, and the case seemed therefore to deserve careful investigation. On inquiry it was found that a Grimsby boat called at the Outer Dowsing on the 15th, the day before the relief took place. The fishing boat had been at sea for a week it was said, and none of the men were ill. A letter was taken to the mate by the crew who boarded the Light-ship and remained on board for some time.

It is evident, therefore, that there were two possible sources of infection—(1) the crew of the fishing boat on the 15th, (2) the crew of the

---

* Public Health, Vol. II.

Trinity House boat. In either case the incubation period must have been less than twenty-four hours.

From what has been said, it appears that the incubation period of influenza may be very short.

It has been suggested that outbreaks of influenza which have been said to occur at sea, may be explained by supposing that the person first seized was infected with the disease before he left port, and that the incubation period in his case was unduly prolonged. This is a matter of pure speculation, and the "fact" which it is intended to explain is too problematical to be worthy of such ingenuity.

CHAPTER XIII.

INFLUENZA IN ANIMALS.

Horses suffer from a disease in which the symptoms resemble those of influenza in human beings, but there is some want of agreement amongst veterinary surgeons as to what disease is meant by, or what diseases are included under, the term influenza. This has been lately commented on by M. Megnin, and M. Veillon. Professor Axe teaches that influenza in man and in the horse are ætiologically distinct, and not capable of passing from one to the other. Professor Fleming has found no evidence to prove that influenza has ever been communicated from horses to mankind or vice versâ. Epidemics of influenza have occurred in man when horses have been free from it, and epizoötics of influenza have prevailed amongst horses when man has been free from the disease. Dogs and cats have suffered from epizoötics in which catarrhal, or influenza symptoms have been predominant. Much difference of opinion has been expressed amongst veterinary surgeons as to the exact relations between human and equine influenza. Mr. Caird believes that influenza may be communicated from horses to cats and to man. M. Auguste Ollivier believes that influenza can be communicated from human beings to cats, and from one cat to another. MM. M. P. Megnin and A. Veillon believe that in 1889, dogs suffered much from a disease analagous to, if not identical with, human influenza. The animals in the gardens of the Royal Zoological Society did not suffer from influenza during the epidemics of 1889, 1890, 1891.

There is no subject of greater interest to medical men, than that of the connexion between the diseases of animals and those of man, and there is no subject in which greater advances in knowledge have been made within recent times. It has long been known of some diseases, and suspected of others, that they may be acquired directly from the lower animals, but it is only within the last few years that the exact method of infection has been certainly known. It has lately been proved, beyond doubt, that in the case of Anthrax at least, the cause of the disease was a living organism which is carried from sheep to man. Quite recently Professor Klein has shown the probability or certainty that diphtheria may be communicated to man from cows and from cats.

In the case of influenza we have still much to learn. It has been shown that epidemics in man have often corresponded in point of time with similar outbreaks in the lower animals, but this has not invariably been the case, and there has hitherto been a singular want of evidence as to direct infection from man to animals or from animals to man.

Horses suffer from a disease in It is a striking fact that horses are subject to a disease very similar to, if not identical with, human influenza. Not only is this the case, but horses suffer also from a catarrhal affection which resembles, in its symptoms, the disease which is commonly called a "feverish" cold, and which, when severe, is not unlike a mild attack of influenza. This disease, which is sometimes called "London fever,"* is always more or less prevalent in the metropolis, and affects chiefly young horses which have recently arrived in town. The London fever is often called influenza, and apparently with as much or as little reason as a feverish cold is called "an influenza cold." It is not usually a severe disease, and has a tendency to run a favourable course.

which the symptoms resemble those seen in the influenza of human beings.

It is evident from what has just been said, that it is difficult or impossible in all cases to determine exactly what disease is indicated when a veterinary surgeon speaks of influenza. MM. M. P. Megnin and A. Veillon† feel this difficulty so strongly that they come to the conclusion that until a more exact definition is found for equine influenza it is useless to draw any deductions from observations on the disease.

There is a want of agreement amongst veterinary authorities as to what disease is meant by the term influenza.

Professor Axe, of the Royal Veterinary College, tells me that his experience leads him to conclude that influenza in man and in the horse are ætiologically distinct and not capable of passing from one to the other.

Professor Axe teaches that influenza in man and in the horse are distinct.

---

* Public Health, June 1891.
† Comptes rendus hebdomadaires des Scéances de la Société de Biologie, q.s., Tom. ii., p. 180.

On this interesting question I have not had sufficient experience to form an opinion. I am glad, therefore, to avail myself of the permission of Professor Fleming, C.B., to publish a communication he kindly sent me on the subject.

Professor Fleming has found no evidence to favour the view that influenza has ever been communicated from horses to mankind or vice versâ.

"The horse is probably the only domesticated animal which suffers from a disease apparently in every respect identical with that which is known as influenza in mankind, and consequently has been given this designation by veterinary surgeons, for though in some rare instances a catarrhal disorder has prevailed among other creatures, and especially dogs, at the same time that influenza was attacking the human species, or horses, or both, yet it has been generally admitted that this was only coincidence, and that there was no relationship between the affections, in a clinical, or even etiological point of view. Neither would there appear to be distinct evidence to support the notion which has been brought forward at times, in favour of an absolute identity in causation between human and equine influenza, and which is perhaps based on the close similarity in symptoms in the two species. For there have been widespread invasions of the disease in the human populations of many countries without horses showing any signs of it: while, on the other hand, most extensive and serious epizoötics have prevailed among the equine species on more than one continent, and yet people were exempt.

Epidemics of influenza have occurred in man when horses have been free from it, and epizoötics of influenza have prevailed amongst horses when there has been no influenza in man.

"Even in the present visitation (1889–1890) we are not in a position to assert that the ordinary seasonal and climatic malady which is usually seen among horses during the late autumn, winter, and early spring months, and which is commonly designated influenza, has been more noticeable in the United Kingdom than in other years. True, from some quarters we hear of more sickness of this description than is ordinarily noted; but these might be looked upon as merely sporadic outbreaks. There has been nothing at all like a universal visitation among horses, such as has been often witnessed, and such as has characterized the human epidemic that has within the last few months manifested itself in such a startling manner on the Continents of Europe and America.

"In the great epizoötics of horse influenza the gravest inconveniences were experienced in large towns and cities by the almost complete cessation of traffic and transit; but nothing of this kind has been reported from any of the countries which have been recently visited by human influenza, and none of my foreign professional correspondents have alluded to horses suffering from any unusual disease.

"In the records of human influenza it is noted that some visitations were preceded or accompanied by a similar affection among horses. In the present epidemic I have not been able to discover that either on the European or the American Continent these animals were affected by any unusual disease. Neither does the history of horse influenza show that an invasion of the malady in one species was at all a certain precedent or accompaniment of the disorder in the other. On the contrary it has repeatedly happened that the equine populations in European countries, and more than once in America, have been generally and most severely visited while mankind has enjoyed complete immunity. (In proof of this I may refer to the history of the malady, given chronologically and descriptively, in my work on Animal Plagues,* in two volumes, published in 1871 and 1882.) Neither is there any evidence that the disorder has ever been communicated from horses to mankind, or *vice versâ*, notwithstanding the great opportunities for such transmission, if such could be effected, which are offered in large horse-establishments and in cavalry regiments.

"When mankind and horses are coincidently affected with influenza, it may therefore be looked upon as a mere casual occurrence, and not as the concomitant result of a common cause. There are presumably two factors operating in the production of the human and the equine disorder, though they must be related to each other in the closest possible manner, as they produce such markedly similar effects, just as do the germs of small-pox, cow-pox, and sheep-pox, and some other human and animal affections.

"The evidence, so far, appears to favour the view entertained by many veterinary surgeons as to the contagiousness of influenza. But it must be acknowledged that there must be an individual predisposition or receptivity, and this, not only for the reception of the virus, but also for the manner in which it will manifest its effects.

Dogs and cats have suffered from epizoötics

"It may be observed that dogs and cats as well as other animals have suffered, though at very long

---

* Op. cit. p. 10.

intervals, from epizoötics, in which catarrhal or influenza symptoms predominated."

with catarrhal symptoms.

At a meeting of the Lincolnshire Veterinary Medical Association, which was held on February 27th, 1890, a discussion arose on the relation between human and equine influenza :—

Some difference of opinion was expressed by members of the Lincolnshire Veterinary Medical Association as to the relation between human and equine influenza.

Professor Pritchard* said :—

"There was one question submitted to me only this morning, and it is worthy of consideration. It is whether there is any chance of influenza being communicated from the horse to the human subject. The question was asked me by Mr. Gresswell, of Louth, and from my own experience I said I did not think it could. It is a fact, no doubt, that the influenza we have been suffering from followed that of the horse, but I do not think there was any connexion between one and the other. Next, Mr. Gresswell stated his experience of an outbreak in his district, and said that not only did he suffer inconvenience from being with his horses, but all his pupils did also. It is a question for our consideration."

Captain Russell said :—

"I am strongly of opinion that if we do not get the same disease in the human subject, we get one similar to it. I could mention several cases, if I had my books here, where I have been called to outbreaks of influenza, and I have not only been seized myself with a severe cold, but I have known

---

* The Veterinarian, Vol. LXIII., Series IV., p. 247.

the men in charge over and over again to be taken in the same way—a generally uncomfortable feeling throughout the whole system, and aches in the muscles and bones, and anything but a desire to work."

Mr. Freer, the President of the meeting, said:—

"I do not know if influenza can be communicated from the animal to man, but I have noticed several men suffering from a cold of a severe form of influenza, as in the horse, at the same time. The whole of the men in the service of Mr. Fernie, master of the Billesdon hounds, have been affected as well as the horses."

Mr. Gooch said:—

"As to the possibility of the disease being communicated from the horse to the human subject, I attended one or two outbreaks on a farm, and out of the twenty-five men on the farm thirteen were laid up with a specific kind of influenza. One of the waggoners attending the horses was not affected, the other was for a week or ten days. I do not think it can be communicated."

Captain Russell explained that he did not mean to say "that the specific influenza we have had this last winter is the one which affects the men in charge of the horses, but I think," he added, "that they may contract a similar complaint from the emanations or issues given off from the animal." It will be noticed there was a great want of direct evidence, and that no definite conclusions were arrived at.

In March last, Mr. J. H. Caird reported some remarkable facts which had come under his notice, and from which he inferred that influenza was communicated from horses to cats and to man. The account of the outbreak is so concise, that I think it best to give it in Mr. Caird's own words:—

Mr. Caird believes that influenza may be communicated from horses to cats and to man.

"At Cairndow, Loch Fyne, on December 24th, 1889, two yearling colts in Mr. J———'s stables were seized with 'strangles,' or what was believed to be that disease. On January 2nd, 1890, two horses, aged respectively five and six years, were seized with influenza; the symptoms were as follows: short cough, profuse nasal discharge, nauseous breath, marked stiffness of joints, disinclination for food or drink; no lung complications; duration of illness about three weeks. I had not the opportunity of seeing these horses, but got the above facts from the owner, a gentleman skilled in the treatment of horses. I had, however, the opportunity of seeing and treating a horse infected in Mr. J———'s stables; this horse was in the infected stables for four hours on January 17th, 1890, and was not in any other infected stables or near any other infected animal. The incubation period was six days, and the period of illness sixteen days. This horse suffered from the following symptoms while under my treatment: short cough, with nasal discharge, at first watery and latterly thick, with a yellowish tint and offensive smell; no glandular swelling of the throat or jaws; he had a weak appearance, staring dry coat, drooping head, dull, sunken eyes, and suffered from loss of appetite, difficulty in urinating, and constipation. For a few days this horse had not lain down in his stall at night. The stethoscope revealed no lung complications. Temperature per rectum 102°; pulse 60 per minute. Mr. J———, the owner of the horses at Cairndow, had three young cats which frequented a hay loft over the stables where the horses were kept while suffering from influenza. These cats were often seen in the stables or about the stables where the horses fed. They were soon seized with sneezing, cough, discharge from eyes and nose, were disinclined to take solid or liquid nourishment, and suffered from severe purging. One died, the others recovered.

" From January 9th, 1890, to February 10th, a dancing class was taught in the hay loft above the stables. This loft was low in the roof, with deficient ventilation, and the infected stables below were in a most insanitary condition. There was direct communication between the mangers out of which the infected horses fed and the hay loft, through which a current of air passed into the loft and contaminated the atmosphere. There were forty pupils at this school, mostly the children of shepherds isolated from each other over an area of about four or five miles from the hay loft where this class was held. There are only about a dozen families in the district, and owing to their pursuits in life and primitive habits they have little communication with the outside world. There was no case of influenza in the district or anywhere near it, as far as I know, before January 1st, 1890. On that date a person, who was in continual contact with the horses, was seized with the disease. On

February 2nd his son and a neighbouring lad were attacked with influenza, complicated with pleuropneumonia. The son described his symptoms as follows :—General malaise for about a week, great depression, sickness and nausea, foul breath, hot and dry skin, alternately with cold sweats, severe pain in loins, frontal headache, deep-seated pain behind the eyeballs, followed by cough and prune-juice expectoration. All these cases were convalescent before I had the opportunity of seeing them, and I cannot describe the symptoms more minutely. All the forty pupils in the dancing class suffered from well-marked influenza except seven. Some had the disease in a mild form, others had the gastro-intestinal type of this disease, one had severe otitis, with foul discharge from the ears. An occasional onlooker at the dancing class, a young lad, took ill on February 7th, and his father the following day; both died on the 15th of the same month. They were ailing more or less for a week before being confined to bed. These were the only two fatal cases. At Lochgoilhead, about nine miles from Cairndow, there were five or six mild sporadic cases after the epidemic at Cairndow, but there were no other cases, to my knowledge, within ten miles of Cairndow."*

Mr. Caird's cases are certainly most striking, and I am surprised that they have not attracted the attention which I think they deserve. Should similar outbreaks be noticed by other observers, current ideas on the question will undoubtedly have to be modified.

---

* Lancet, 1891, Vol. I., p. 741.

M. Auguste Ollivier believes that influenza can be communicated from human beings to cats, and from one cat to another.

At a meeting of l'Académie de Médecine (Paris) on January 28th, 1889, M. Auguste Ollivier[*] related a case which seemed to him to prove that influenza could be communicated to a cat by matter expectorated by a human subject.

A lady had a violent attack of influenza, which was accompanied by the expectoration of much viscid mucous. As she was extremely feeble, and took food with difficulty, her doctor advised her to suck pieces of meat. She extruded from her mouth several morsels which she had chewed, when a cat seized upon them and swallowed them. Three or four days later the animal died, having suffered from symptoms of influenza, namely, from cough, vomiting, dyspnœa, and loss of flesh.

M. Ollivier referred, at the same meeting, to a communication he had made to the Société de Biologie as long ago as 1875. This was as follows:—In 1868, during the presence of an epidemic of influenza, a sick cat took refuge with some good people who were strong friends to the feline race, for they had already five cats, all at that time in a good state of health. The new comer was not driven away, and he eat at the same mess as the others, letting fall into the food particles of mucous which ran down from his nose and mouth. For six days the six cats lived thus. After the lapse of this time the newcomer succumbed, exhausted by cough and vomiting. M. Ollivier made an autopsy with the greatest care, and found well marked lesions identical with those seen in the organs of

---

[*] Bulletin de l'Académie de Médecine, 3 S., Tome xxiii., pp. 118, 119.

persons who have died of influenza, and notably patches of pneumonia which M. Ranvier recognised to be analogous to those which are seen in the broncho-pneumonia of children.

On the day on which the stranger-cat died it was seen that two of the cats of the house had influenza in their turn. The next day and the day after the three other cats were similarly affected. In each case the disease recalled symptom by symptom the malady of the first cat, and became fatal in eight or nine days in four of these animals.

The autopsies gave the same results.

MM. M. P. Megnin and A. Veillon brought before la Société de Biologie* some observations they had made concerning the diseases of dogs, and they gave it as their opinion that in certain kennels an infectious malady which had a great analogy to human influenza was very prevalent in 1889. One of these gentlemen investigated an outbreak at Chantilly, where the majority of a pack of one hundred and twenty dogs suffered, a second in Maine-et-Loire in a pack of fifty beagles, and a third in l'Aveyron amongst some greyhounds.

At Chantilly and l'Aveyron especially, the symptoms of human influenza were most marked, and the dogs suffered from lachrymation, swollen eyes, a distressing cough, and general prostration.

M. M. P. Megnin and M. A. Veillon believe that in 1889 dogs suffered from a disease analogous to human influenza.

During the spring of 1891 I saw many cats which appeared to me to be suffering from a

The animals in the Royal Zoological Society's

---

* Comptes rendus hebdomadaires des Séances de la Société de Biologie, q. s., Tom. ii., p. 180.

U p. 1501.

Gardens did not suffer from influenza during the epidemics of 1889, 1890, 1891.

disease analogous to human influenza, and I observed in some cases an amount of prostration which amounted almost to paralysis. Mr. Bartlett, of the Zoological Society, tells me that during the epidemics of influenza in 1889, 1890, and 1891 there was no unusual mortality amongst the animals in the Society's Gardens, and that he has no reason to suppose that any of them suffered from it or any like disease.

CHAPTER XIV.

THE NOTIFICATION OF INFLUENZA SHOULD BE COMPULSORY.

The Three Estates of this Realm have in their wisdom made laws for the protection of the people against infectious diseases.

The "Infectious Disease (Notification) Act, 1889,"* was a great advance in this direction. Section 6 of this Act is as follows:—

"In this Act the expression 'infectious disease' to which this Act applies means any of the following diseases, namely, small-pox, cholera, diphtheria, membranous croup, erysipelas, the disease known as scarlatina or scarlet fever, and the fevers known by any of the following names, typhus, typhoid, enteric, relapsing, continued, or puerperal, and includes as respects any particular district any infectious disease to which this Act has been applied by the local authority of the district in manner provided by this Act."

Influenza is not included amongst the diseases to which the Act always applies, but the local authority of any district may under conditions mentioned in

---

* 52 & 53 Vict. c. 72. *See* Appendix.

section 7 apply the provisions of the Act to any infectious disease.

Now, there are Medical Officers of Health who apparently do not know that influenza is infectious, and it can hardly be supposed that local authorities are better informed. It follows from this that the provisions of the Act will not be universally carried out in the case of influenza, so long as local authorities have the right to use, or to neglect to use, the powers conferred on them.

It is, therefore, to be hoped that influenza will, by a short Act of Parliament, be placed in the position to which it is justly entitled amongst the infectious diseases for which notification is compulsory.

Should these notes help to accomplish so desirable an end they will not have been written in vain.

APPENDIX.

Infectious Disease (Notification) Act, 1889.
[52 & 53 Vict. Ch. 72.]

Arrangement of Sections.

Section.
1. Short title.
2. Extent of Act.
3. Notification of infectious disease.
4. As to forms and case of several medical practitioners.
5. Adoption of Act in urban or rural district.
6. Definition of infectious disease.
7. Power to local authority to extend definition of infectious disease.
8. Notices and certificates.
9. Expenses.
10. Repayment of expenses in London as expenses of managers of asylum district.
11. Non-disqualification of medical officer by receipt of fees.
12. Application of Act to Woolwich.
13. Application of Act to vessels, tents, &c.
14. Saving for local Act.
15. Exemption of Crown buildings.
16. Definitions.
17. Application of Act to Scotland.
18. Application of Act to Ireland.

CHAPTER 72.

An Act to provide for the Notification of Infectious Disease to Local Authorities.

[30th August 1889.]

BE it enacted by the Queen's most Excellent Majesty, by and with the advice and consent of the Lords Spiritual and Temporal, and Commons, in this present Parliament assembled, and by the authority of the same, as follows :—

Short title.

1. This Act may be cited as the Infectious Disease (Notification) Act, 1889.

Extent of Act.

2. This Act shall extend—
   (*a*) to every London district after the expiration of two months from the passing of this Act, and
   (*b*) to any urban, rural, or port sanitary district after the adoption thereof.

Notification of infectious disease.

3.—(1.) Where an inmate of any building used for human habitation within a district to which this Act extends is suffering from an infectious disease to which this Act applies, then, unless such building is a hospital in which persons suffering from an infectious disease are received, the following provisions shall have effect, that is to say :—
   (*a*) the head of the family to which such inmate (in this Act referred to as the patient) belongs, and in his default the nearest relatives of the patient present in the building or being in attendance on the patient, and in default of such relatives every person in charge of or in attendance on the patient, and in default of

any such person the occupier of the building shall, as soon as he becomes aware that the patient is suffering from an infectious disease to which this Act applies, send notice thereof to the medical officer of health of the district :

(b.) every medical practitioner attending on or called in to visit the patient shall forthwith, on becoming aware that the patient is suffering from an infectious disease to which this Act applies, send to the medical officer of health for the district a certificate stating the name of the patient, the situation of the building, and the infectious disease from which, in the opinion of such medical practitioner, the patient is suffering.

(2.) Every person required by this section to give a notice or certificate who fails to give the same, shall be liable on summary conviction in manner provided by the Summary Jurisdiction Acts to a fine not exceeding forty shillings ;

Provided that if a person is not required to give notice in the first instance, but only in default of some other person, he shall not be liable to any fine if he satisfies the court that he had reasonable cause to suppose that the notice had been duly given.

4.—(1.) The Local Government Board may from time to time prescribe forms for the purpose of certificates under this Act, and any forms so prescribed shall be used in all cases to which they apply. As to forms and case of several medical practitioners.

(2.) The local authority shall gratuitously supply forms of certificate to any medical practitioner residing or practising in their district who applies for the same, and shall pay to every medical practitioner for each certificate duly sent by him in accordance with this Act a fee of two shillings and sixpence if the case occurs in his private practice, and of one shilling if the case occurs in his practice as medical officer of any public body or institution.

(3.) Where in any district of a local authority there are two or more medical officers of health of such authority a certificate under this Act shall be given to such one of those officers as has charge of the area in which is the patient referred to in the certificate, or to such other of those officers as the local authority may from time to time direct.

<small>Adoption of Act in urban or rural district.</small>

5.—(1.) The local authority of any urban, rural, or port sanitary district may adopt this Act by a resolution passed at a meeting of such authority; and fourteen clear days at least before such meeting special notice of the meeting, and of the intention to propose such resolution, shall be given to every member of the local authority, and the notice shall be deemed to have been duly given to a member if it is either:

(a.) given in the mode in which notices to attend meetings of the local authority are usually given, or

(b) where there is no such mode, then signed by the clerk of the local authority and delivered to the member or left at his usual or last known place of abode in England, or forwarded by post in a prepaid letter addressed to the member at his usual or last known place of abode in England.

(2.) A resolution adopting this Act shall be published by advertisement in a local newspaper, and by handbills, and otherwise in such manner as the local authority think sufficient for giving notice thereof to all persons interested, and shall come into operation at such time, not less than one month after the first publication of the advertisement of the resolution as the local authority may fix, and upon its coming into operation this Act shall extend to the district.

(3.) A copy of the resolution shall be sent to the Local Government Board when it is published.

<small>Definition of infectious disease.</small>

6. In this Act the expression "infectious disease to which this Act applies" means any of the following diseases, namely, small-pox, cholera, diphtheria, membranous croup, erysipelas, the disease known as scarlatina or scarlet fever, and the fevers known by any of the following names, typhus, typhoid, enteric, relapsing, continued, or puerperal, and includes as respects any particular district any infectious disease to which this Act has been applied by the local authority of the district in manner provided by this Act.

7.—(1.) The local authority of any district to which this Act extends may, from time to time, by a resolution passed at a meeting of such authority where the like special notice of the meeting and of the intention to propose the resolution has been given as is required in the case of a meeting held for adopting this Act, order that this Act shall apply in their district to any infectious disease other than a disease specifically mentioned in this Act.

Power to local authority to extend definition of infectious disease.

(2.) Any such order may be permanent or temporary, and, if temporary, the period during which it is to continue in force shall be specified therein, and any such order may be revoked or varied by the local authority which made the same.

(3.) An order under this section and the revocation and variation of any such order shall not be of any validity until approved by the Local Government Board.

(4.) When it is so approved, the local authority shall give public notice thereof by advertisement in a local newspaper and by handbills, and otherwise in such manner as the local authority think sufficient for giving information to all persons interested. They shall also send a copy thereof to each registered medical practitioner whom, after due inquiry, they ascertain to be residing or practising in their district.

(5.) The said order shall come into operation at such date not earlier than one week after the publication of the first advertisement of the approved order as the local authority may fix, and upon such order coming into operation, and during the continuance thereof, an infectious disease mentioned in such order shall, within the district of the authority, be an infectious disease to which this Act applies.

(6.) In the case of emergency three clear days' notice under this section shall be sufficient, and the resolution shall declare the cause of such emergency and shall be for a temporary order, and a copy thereof shall be forthwith sent to the Local Government Board and advertised, and the order shall come into operation at the expiration of one week from the date of such advertisement, but unless approved by the Local Government Board shall cease to be in force at the expiration of one month after it is passed, or any earlier date fixed by the Local Government Board.

(7.) The approval of the Local Government Board shall be conclusive evidence that the case was one of emergency.

Notices and certificates.

8.—(1.) A notice or certificate for the purposes of this Act shall be in writing or print, or partly in writing and partly in print; and for the purposes of this Act the expression "print" includes any mechanical mode of reproducing words.

(2.) A notice or certificate to be sent to a medical officer of health in pursuance of this Act may be sent by being delivered to the officer or being left at his office or residence, or may be sent by post addressed to him at his office or at his residence.

Expenses.

9. Any expenses incurred by a local authority in the execution of this Act shall be paid as part of the expenses of such authority in the execution of the Acts relating to public health and in the case of a rural authority shall be general expenses.

Repayment of expenses in London as expenses of managers of asylum district.

10. Where a medical officer of health receives in pursuance of this Act a certificate of a medical practitioner relating to a patient within the metropolitan asylum district, he shall within twelve hours after such receipt forward a copy thereof to the managers of that district, and those managers shall repay to the local authority the amounts paid by that authority in respect of those certificates of which copies have been sent to the managers as required by this section, and shall repay those amounts out of the fund out of which the general expenses of the managers are paid. The managers shall send weekly to the London County Council such return of the infectious diseases of which they receive certificates in pursuance of this Act as the London County Council from time to time require.

Non-disqualification of medical officer by receipt of fees.

11. A payment made to any medical practitioner in pursuance of this Act shall not disqualify that practitioner for serving as member of the council of any county or borough, or as member of a sanitary authority, or as guardian of a union, or in any municipal or parochial office.

Where a medical practitioner attending on a patient is himself the medical officer of health of the district, he shall be entitled to the fee to which he would be entitled if he were not such medical officer.

12. This Act shall apply to the Local Board of Woolwich in like manner as if it were a vestry under the Metropolis Management Act, 1855, and that board shall appoint and pay a medical officer of health, and all enactments relating to medical officers of health within the administrative county of London shall apply to the medical officer of health of Woolwich.

Application of Act to Woolwich. 18 & 19 Vict. c. 120.

13.—(1.) The provisions of this Act shall apply to every ship, vessel, boat, tent, van, shed, or similar structure used for human habitation, in like manner as nearly as may be as if it were a building.

Application of Act to vessels, tents, &c.

(2.) A ship, vessel, or boat, lying in any river, harbour, or other water not within the district of any local authority within the meaning of this Act shall be deemed for the purposes of this Act to be within the district of such local authority as may be fixed by the Local Government Board, and where no local authority has been fixed, then of the local authority of the district which nearest adjoins the place where such ship, vessel, or boat is lying.

(3.) This section shall not apply to any ship, vessel, or boat belonging to any foreign Government.

14. Where this Act is put in force in any district in which there is a local Act for the like purpose as this Act, the enactments of such local Act, so far as they relate to that purpose, shall cease to be in operation.

Saving for local Act.

15. Nothing in this Act shall extend to any building, ship, vessel, boat, tent, van, shed, or similar structure belonging to Her Majesty the Queen, or to any inmate thereof.

Exemption of Crown buildings.

16. In this Act—

Definitions.

The expression "local authority" means each of the following authorities; that is to say,—
(a) the Commissioners of Sewers in the City of London;

(b) the vestry under the Metropolis Management Act, 1855, of a parish in Schedule A, and the district board of a district in Schedule B to the Metropolis Management Act, 1855, as amended by the Metropolis Management Amendment Act 1855, and the Metropolis Management (Battersea and Westminster) Act, 1887;

18 & 19 Vict. c. 120.
48 & 49 Vict. c. 33.
50 & 51 Vict. c. 17.

(c) an urban or rural sanitary authority in England within the meaning of the Public Health Acts; and

(d) the port sanitary authority of any port sanitary district in England.

The expression "London district" means the City of London or the parish or district mentioned in Schedule A or Schedule B of the Metropolis Management Act, 1855, for which a local authority is elected:

The expression "urban or rural district" means the district for which any such urban or rural sanitary authority is elected:

The expression "port sanitary district" means the port sanitary district of London and any port or part of a port for which a port sanitary authority has been constituted under the Public Health Acts, and any such port sanitary district shall form no part, for the purposes of this Act, of any urban or rural district:

The expression "occupier" includes a person having the charge, management, or control of a building, or of the part of a building in which the patient is, and in the case of a house the whole of which is let out in separate tenements, or in the case of a lodging-house the whole of which is let to lodgers, the person receiving the rent payable by the tenants or lodgers either as his own account or as the agent of another person, and in the case of a ship, vessel, or boat, the master or other person in charge thereof.

Application of Act to Scotland.

17. In the application of this Act to Scotland—

The expression "Local Government Board" shall mean Board of Supervision:

The expression "Summary Jurisdiction Acts" shall mean the Summary Jurisdiction (Scotland) Acts 1864 and 1881, and any Act amending the same:

The expression "local authority" shall mean the local authority as defined by the Public Health (Scotland) Act, 1867, and any Act amending the same:

The expression "England" in section five shall mean Scotland:

The powers contained in this Act shall be in addition to and not in lieu of any powers existing in any local authority by virtue of any general or local Act.

18. This Act shall apply to Ireland, with the following modifications:

Application of Act to Ireland.

(1.) In this Act, unless the context otherwise requires—

The expression "Local Government Board" means the Local Government Board for Ireland:

The expression "local authority" means an urban or rural sanitary authority within the meaning of the Public Health (Ireland) Act, 1878:

41 & 42 Vict. c. 52.

The word "district" means urban sanitary district or rural sanitary district, as the case may be, within the meaning of the said Act:

The expression "clerk of the local authority" includes, in the case of an urban sanitary authority, town clerk and secretary:

(2.) References to a place of abode in England shall be construed to refer to a place of abode in Ireland.

(3.) Offences under this Act may be prosecuted, and fines under this Act may be recovered, in manner directed by the Summary Jurisdiction Acts, before a court of summary jurisdiction constituted in the manner mentioned in the two hundred and forty-ninth section of the Public Health (Ireland) Act, 1878.

41 & 42 Vict. c. 52.

INDEX.

	Page
Adams, Mr. Alexander	39
Among the Mongols	12
Animal Plagues, their History, Nature, and Prevention	10
Animals, influenza in	117–130
An Epitome of the Report of the Medical Officers to the Chinese Imperial Maritime Customs Service from 1871 to 1882	12
Annals of Influenza	38
Antony, Dr.	59
Asia, Central, origin in	11
A system of Medicine	7
A system of Practical Medicine, by American Authorities	107
"Atlas" East Indiaman, reported sudden outbreak on the (1780)	31
A Treatise on the Principles and Practice of Medicine	16
A Treatise on the Theory and Practice of Medicine	7
Axe, Professor	119
Bacteriologists, promising field for, at Swatow	13
Bacteriology, importance of	16–17
Bäer, views of	24
Barker, Mr. W. J. Townsend, of Churchingford	81–82
Barnardo's Homes, epidemic at Dr.	67
Barth, M.	59
Bartlett, F.Z.S., Mr.	130
Bäumber, Professor Chr.	52
Bayswater, epidemic at (1889)	67
Becher, views of	24
Bengal affected (1870)	31
Berkshire, spread in rural parts of	80
Berlin, epidemic at (1889)	52
Bernard, Mr., of Jersey	105
Berry Head, reported sudden outbreak on board ship off (1833)	28
Birmingham, isolated cases at, before epidemic	73
Blake, Dr., of Great Yarmouth	83, 115
Blyth, Professor Wynter	114

	Page
Bokhara, epidemic at (1889)	47–50
Boobbyer, Dr., of Nottingham	77
Bordone, Dr.	56, 111
Bouchard, Professor	25, 26, 56, 57, 59
"Bretagne," spread by contagion on the training ship	60–61
Bristowe, Dr. John Syer	7
British Medical Journal, The	18, 39, 71, 73, 74, 75, 80, 83, 92
Broadwood's Factory, outbreak at (1889–1890)	69, 70
Broussais, l'Hôpital, spread by contagion in	59
Buchanan, Dr.	64
Bulletin de l'Académie de Médecine, Paris	56, 59, 128
Bulletin et Memoirs de la Société Médicale des Hôpitaux de Paris	59, 111
Bulletin Médical	11
Caird, Mr. J. H.	125–127
Canterbury, influenza at	70
Cardiff, isolated cases at, before epidemic	74
Cats, influenza in	122, 125–129
Causes, various, assigned	9–17
Chantilly, dogs affected at	129
Charlottenbourg, nuns escape influenza at	91, 92
Chelmsford, isolated cases at, before epidemic	71
Chester, spread by contagion (1775 and 1782)	41, 42
China affected	12, 13
Churchingford, imported into, from neighbouring towns (1890)	82–88
Churchingford, imported into, from Paris (1889)	81–88
Churchman's Magazine, The	9
Clark, Dr., of Newcastle	38
Classification, clinical	1
„ pathological	2
Clemow, Dr. Frank	4, 34, 84
Clifton, influenza at	83
Colin, M.	25, 61
Collingridge, Dr.	114
Commons, House of, outbreak amongst members of the	84
Comptes rendus hebdomaidaires des Séances de la Société de Biologie	119–129
Contagion, nature of, evidence in favour of	33–36
Coromandel, influenza on the coast of	31
"Correo," incubation period of a case on board the	114
Crediton, imported into, from Exeter (1803)	44
Crookshank, Professor	17
Cullen, Dr.	27
Currie, Dr., of Liverpool	32

	Page
Daly, Dr. E. O., of Hull	39–83
Definition of influenza	6–8
,, ,, by an anonymous writer in "The Times"	8
,, ,, by Dr. Bristowe	7
,, ,, by Dr. Parkes	6
Delépine, Dr. Sheridan	69, 70, 112, 113
Derbyshire, spread in the rural parts of	80
Des Déserts, Dr. Danguy	60, 105
D'Hoste, Dr.	59
Docks, influenza at the	67
Dogs, influenza in	122–129
Dorchester, spread by contagion in a country house near	75
Downs, possible infection from a pilot taken on board in the	114
,, return of affected fleet to the (1782)	30
Edinburgh affected	81
Elkins, Mr. Frank A.	92
"Epidemic catarrh," synonym for influenza	31
Feldburg, imported into, from Freiburg	52
Fleming, Professor George, C.B.	10, 120–3
Flint, Dr. Austin	16
Freer, Mr.	124
Freiburg, spread by contagion in hospital at	53
French names for influenza	3
Frontignan, incubation period in case at	111
,, imported into, from Paris	56
Fürbinger	24
Gentleman's Magazine, The	4
German names for influenza	3
Germ theory of disease, the	9, 13–17
Gilmour, Mr.	12
"Goliah," reported sudden outbreak on board the (1782)	30
Gooch, Mr.	124
Gordon, M.D., Surgeon-General C. A.	12
Grasset, M. le Professeur J.	25, 54
Gray, Dr. Edward	19, 31, 32, 33, 39, 89, 102, 104
Guiteras, Dr. John	107
Haileybury College, epidemics at (1889–1890)	85, 86
Haileybury, outbreak at	5
Hamilton, Dr.	40, 41, 45
Hammersmith, epidemic at	67
Handbook of Geographical and Historical Pathology	20

	Page
Hankin, the Rev. Daniel B.	9
Hart, Mr. Horace	71-73
Haygarth, Dr. John	41, 42
Henry, Mr., of Manchester	32
Hereford Asylum, inmates escaped influenza in (1803)	91
Hirsch, Professor August	20-24, 27
Holland, Sir Henry	17, 27
Horses, influenza in	118-127
House of Commons, outbreak amongst members of the	84
Howe, Lord, reported sudden outbreak on board ships under his command (1782)	30
Hugo, Mr., of Crediton	44
Hull, imported into, by sailors	39, 83
Huxham, Dr.	6, 37
Imber, Mr. Naphtali Herz	11
Influenza Epidemic; a Visitation from God, The	9
Influenza, introduction of name to England	4
,, origin of name	4
,, reasons for adopting this name	5
Inverness, affected	80
Ipswich, spread by contagion in (1782)	40
Ireland, Report of Local Government Board for	26
Jail, Birmingham	92
,, Bodmin	93
,, Canterbury	93
,, Exeter	93
,, Ipswich	94
,, Kendal	94
,, Knutsford	94
,, Leicester	94
,, Lewes	95
,, Liverpool	95
,, Maidstone	95
,, Norwich	95
,, Portsmouth	95
,, Preston	95
,, Ruthin	95
,, St. Albans	95
Jails, Scottish	99, 100, 101
Jail, Strangeways	96
,, Wandsworth	96
,, Winchester	97
,, Worcester (1803)	91
Jewish Standard	11

	Page
Kempenfelt, Admiral, reported sudden outbreak on ships under the command of (1782)	30
Keynsham, outbreak at	83
Kiakta, prevalent in neighbourhood of	12
Kilmersham, outbreak at	81
Klein, Professor	14–17
"Kurdaikis," synonym for influenza	11
Lancet, The	28, 52, 64, 65, 74, 77, 105, 106, 127
Latin names for	2, 3
L'Aveyron, influenza in dogs at	129
Leçons sur la Grippe de l'hiver 1889–1890	25, 54
Lectures on the Principles and Practice of Physic	26, 28
Lincolnshire Veterinary Medical Association, meeting of the	123, 124
Liverpool, isolated cases in (1890)	74
London, epidemic of 1782	45, 46
,, epidemic of 1889, 1890	64–70
,, reported sudden outbreak in (1833)	28
"London fever" in horses	119
Lunel, incubation period in which infection was derived at	112
,, spread by contagion at	58
Lush, Dr. William	75
Maclagan, Sir Douglas, M.D.	99–101
Manchester, epidemic at (1889, 1890)	78–80
Maine-et-Loire, influenza in dogs in	129
Mease, Dr.	104
Medical and Physical Journal	91
Medical Communications	19, 31, 38, 89, 102, 104
Medical Knowledge, Society for Promoting	31
Medical Notes and Reflections	17, 27
Medical Transactions, published by the College of Physicians in London	30, 45
Megnin, M.	119, 129
Memoirs of the Medical Society of London	40, 45, 91
Middlesbrough, epidemic at	80
Mongols, influenza amongst the	12
Montbéliard, imported into, from surrounding towns	56
Morningside Asylum, outbreak at	92
Munro, Dr.	35
Newcastle-on-Tyne, probably imported from Gothenburg to	73
New York, death rate during epidemic in 1890	69
,, imported from	83
Nomenclature of disease	1

			Page
Nomenclature of influenza			1–5
,, ,, founded in fancy			4, 5
,, ,, ,, supposed origin			3
,, ,, ,, symptoms and peculiarities			2, 3
Northampton, outbreak at			73
Northern Chronicle, The			80
Norwich, spread by contagion (1792)			41
Nottingham, outbreak at			77, 78
Ogle, Dr. John W.			52
Ollivier, M. Auguste			128, 129
Origin in Central Asia			11, 12
,, China			12, 13
Outer Dowsing Light-ship, case of the mate of			115, 116
Oxford University Press, outbreak at the			71–73
Paris, epidemic in (1889)			55
Parkes, Dr., definition of influenza by			7
Parsons, Dr.			115
Pearson, Dr.			7
Philadelphia Times, The			107
Portsmouth, reported sudden outbreak at			29
Post Office, London General, outbreak at			67
Practitioner, The			69, 112
Prager Medicinische Wochenschrift			108
"Princess Amelia," reported sudden outbreak on board the (1782)			30
Prisons. *See* Jails.			
Prisoners and influenza			89–102
Prisons, Commissioners for Scotland, Twelfth Annual Report			99
Pritchard, Professor			123
Proust, M. le Professeur			37, 59
Public Health			4, 34, 75, 83, 84, 115, 119
Rabbis, views of the			11
Renvers, views of			24
Report of the Commissioners of Prisons, The Thirteenth			93–96
Reygate (Reigate) Workhouse, paupers escaped influenza at			91
"Ripon," reported sudden outbreak on board the			30
Robertson, Mr. George M.			92
Russell, Captain			124
St. George's Hospital, casualty patients at			68
"S. Germain," spread by contagion on board the			59

	Page
St. Ives, outbreak at	80
St. James's Gazette	66
St. Petersburgh, epidemic at (1889)	50, 51
Schauta, Professor	108, 109
Schnirer, views of	24
Semaine Médicale	91, 105
Sheen, Dr. A.	74
Sheffield, cases imported into, 1890	75
,, ,, ,, 1891	83
,, epidemic of 1890	75, 76
,, ,, 1891	83, 84
Shelly, Dr., of Hertford	65, 85, 86
Shields, imported into, by sailors (1782)	39
Ships at sea, reported outbreaks on board	28–32
Short, Dr.	7
Simmons, Dr. Samuel Foart	40
Société de Biologie	129
Spadaccini, Mr. Henry	114
Squire, Dr. William	28
"Stag," reported sudden outbreak on the	28, 29
Suffolk, spread by contagion in	40
Swatow, epidemic at	12, 13
Tatham, Dr. John, of Manchester	78–8), 96
Taylor, Dr. G. C., of Trowbridge	81
Taylor, Dr., of Cardiff	74
Teissier, M.	11
Thorne, Dr. Bezley	106
Thompson, Dr. Theodore, of Sheffield	75, 83
Thompson, Dr. Theophilus	6, 10, 38, 41
Times, The	8, 71–3
Trowbridge, epidemic at	81
Tueffart, Dr.	56, 112
Ueber die Influenza von 1889 *und* 1890	52
Urga (Mongolia), prevalent in the neighbourhood of	12
Universal Review, The	14
Unsere Zeit	47
Val de Grâce, spread by contagion in the hospital of	59, 111
Védel, Dr., of Lunel	58, 112
Veillon, M.	119, 129
Vergèze, imported into, from Lunel	58
Veterinarian, The	123
Vic, imported into, from Frontignan	56
Vienna, death rate during the epidemic of 1889	52

	Page
Wales, North, epidemic in rural parts of	81
Watson, Dr., of Westbourne Grove	66
Watson, Sir Thomas	26–29, 34
Westbourne Grove, outbreak in	65, 66
West India Dock, crew of " Correo " affected in	114
Westport (county Mayo), imported into, by sailors	39
White, Dr. J. William	107
Wilks, Dr. Samuel, extract from letter from	4, 5
Wimborne, epidemic at	80
Zoological Society, animals in the gardens of the Royal, escaped influenza	136

www.ingramcontent.com/pod-product-compliance
Lightning Source LLC
Chambersburg PA
CBHW020247170426
43202CB00008B/266